Elementary
Regression Modeling

To Elaine.

SAGE was founded in 1965 by Sara Miller McCune to support the dissemination of usable knowledge by publishing innovative and high-quality research and teaching content. Today, we publish over 900 journals, including those of more than 400 learned societies, more than 800 new books per year, and a growing range of library products including archives, data, case studies, reports, and video. SAGE remains majority-owned by our founder, and after Sara's lifetime will become owned by a charitable trust that secures our continued independence.

Los Angeles | London | New Delhi | Singapore | Washington DC | Melbourne

Elementary Regression Modeling

A Discrete Approach

Roger A. Wojtkiewicz
Ball State University

Los Angeles | London | New Delhi
Singapore | Washington DC | Melbourne

FOR INFORMATION:

SAGE Publications, Inc.
2455 Teller Road
Thousand Oaks, California 91320
E-mail: order@sagepub.com

SAGE Publications Ltd.
1 Oliver's Yard
55 City Road
London EC1Y 1SP
United Kingdom

SAGE Publications India Pvt. Ltd.
B 1/I 1 Mohan Cooperative Industrial Area
Mathura Road, New Delhi 110 044
India

SAGE Publications Asia-Pacific Pte. Ltd.
3 Church Street
#10-04 Samsung Hub
Singapore 049483

Acquisitions Editor: Leah Fargotstein
Editorial Assistant: Yvonne McDuffee
eLearning Editor: Nicole Mangona
Production Editor: Kelly DeRosa
Copy Editor: Sheree Van Vreede
Typesetter: C&M Digitals (P) Ltd
Proofreader: Wendy Jo Dymond
Indexer: Jeanne Busemeyer
Cover Designer: Anupama Krishnan
Marketing Manager: Susannah Goldes

Printed in the United States of America

Library of Congress Cataloging-in-Publication Data

Names: Wojtkiewicz, Roger A., author.

Title: Elementary regression modeling : a discrete approach / by Roger Wojtkiewicz, Ball State University.

Description: Los Angeles : SAGE, [2017] | Includes bibliographical references and index.

Identifiers: LCCN 2015046051 | ISBN 9781506303475 (pbk. : alk. paper)

Subjects: LCSH: Social sciences—Statistical methods. | Regression analysis.

Classification: LCC HA31.3. W65 2017 | DDC 300.1/519536—dc23 LC record available at http://lccn.loc.gov/2015046051

This book is printed on acid-free paper.

16 17 18 19 20 10 9 8 7 6 5 4 3 2 1

Brief Contents

Access helpful study tools and resources—all in one place!

The data and statistical program code for this book are also available on the SAGE website at study.sagepub.com/wojtkiewicz.

Detailed Contents

Access helpful study tools and resources—all in one place!

The data and statistical program code for this book are also available on the SAGE website at study.sagepub.com/wojtkiewicz.

Preface

Purpose of the Book

Social scientists use regression analysis extensively to analyze quantitative data, and as a result, there are many books available about regression analysis. My book makes a unique contribution by using a discrete approach to discuss regression modeling. I wrote my book with two key purposes in mind. One was to present an approach to understanding regression analysis that was more straightforward and understandable than the usual approach to explaining regression analysis. Over the course of many years teaching regression analysis to students, I have developed this discrete approach for explaining regression analysis. My discrete approach views the coefficients for the independent variables in a regression equation as capturing group differences on the dependent variable. Therefore, this approach is in contrast to the usual approach to explaining regression analysis that involves continuous independent and dependent variables and estimating a line through a cloud of points. The discrete approach builds on simpler discrete analyses in statistics that use chi-square and t tests. My students have found this approach to be easily understandable.

My second purpose was to provide a source on how to do regression modeling in social science research. I define regression modeling in my book as the use of one or more regression equations to examine a particular coefficient in greater depth. To do that, we need to address research hypotheses. The many books on regression analysis cover well the standard continuous, cloud-of-points approach to explaining regression analysis, technical issues in regard to estimating appropriate regression models given the data at hand, and advanced methods for doing regression analysis. However, what has not been addressed thoroughly is how to use a series of multiple regression equations to examine the contributions of control variables, different approaches to estimating interaction effects, and methods for examining linearity in regression effects.

Many social science graduate programs include a two-semester sequence in statistics with the second semester focusing on regression analysis. In my experience, I have found that students after taking such a sequence understand what regression is, how to use a statistical program to estimate regression coefficients, and what are the technical issues that may cause problems in regressions. However, I also find after this course that many students are still not equipped to answer research questions by using regression analysis. My purpose in writing this book is to provide tools involving control modeling, interaction modeling, and using splines to model linearity that students and researchers can apply to address social science research hypotheses.

Intended Audience

There are two primary audiences for this book. First, this book will be a valuable resource for instructors and students in the second-semester regression course that is taught in many social science graduate programs. I envision this book being a second required text for such a course. The primary required book will be one of the standard texts on regression that covers basic regression analysis, technical issues, and more advanced models. My book would then allow instructors to build on the standard text by adding course content on how to apply regression modeling to address research hypotheses. Using the chapters in my book on control modeling, modeling interactions, using splines to model linearity, and research hypotheses will better prepare students for theses, dissertations, research projects, and contributing to faculty research projects,

The second intended audience for my book is experienced researchers who use regression analysis on an ongoing basis. Although the first three chapters of the book present the discrete approach to understanding regression analysis and deal with the basics of regression analysis, experienced researchers may find this alternative way of thinking about regression analysis informative to their overall understanding of regression. In addition, those who teach undergraduate statistics will find the discrete approach to be a much more accessible way for students to understand regression analysis than the explanations found in most undergraduate statistics books. The chapters on regression modeling concepts, control modeling, modeling interactions, and modeling linearity with splines present more sophisticated ideas. The chapters deal with these issues in an innovative way and in a more in-depth manner than found in other references on regression analysis. I envision my book as the kind that researchers will keep on their bookshelves and often refer to when working on regression modeling issues.

Unique Contributions

My book makes several contributions to understanding regression and regression modeling:

Discrete approach: A key contribution of the book is the discrete approach to understanding regression analysis. The discrete approach builds on simple differences between groups to explain regression and regression modeling. Although the discrete approach is fully consistent with the common linear, cloud-of-points approach to explaining regression, the discrete approach is much more immediately understandable.

Means and log odds: The book starts out by looking at means and log odds and then works up to dummy variable regression. This is a simple-to-understand way to approach regression as opposed to the usual abstract starting point that involves estimating a least-squares line through a cloud of points. The book also demystifies logistic regression by showing how the log odds is related to percentages and proportions and how logistic regression with a dummy independent variable just estimates a difference in log odds between two groups.

Linearity as jumps between groups: The discrete approach starts with dummy variables and then moves to understanding interval or continuous variables. The book shows that a coefficient for an interval variable can be viewed as an equality constraint on the jumps in the mean between groups as one goes from lower groups on the independent variable to higher groups.

Unit vector, nestedness, higher order differences, constraints: The book explicitly and in detail addresses issues that receive little or no direct coverage in other books on regression analysis. Most books on regression analysis do not give much attention to the unit vector, but as my book shows, understanding the role of the unit vector is important to understanding what the other coefficients mean in regression analysis. In the background of any discussion of regression analysis are the concepts of nestedness, higher order differences, and constraints. My book brings these concepts to the forefront and shows how they apply in different modeling situations.

Control modeling: Starting with a model with a smaller number of variables and building to larger models with more variables is one of the most, if not the most, commonly used modeling approach in social science research. Surprisingly, other books on regression give little attention to this issue. My book brings to the forefront the underlying issue in control modeling, which is the correlation between independent variables. The book provides alternative modeling approaches for dealing with the problem of how order of entry of independent variables affects the nature of the resulting explanation. The book introduces new terminology for control models: one-at-a-time with no controls, one-at-a-time with controls, step model, and hybrid model. The book also shows how the method of demographic standardization can be used to understand what actually happens when a control variable is added to a model.

Modeling interactions: Creating interaction variables is a simple task, merely multiply two variables. However, interpreting the meaning of an interaction coefficient in a regression model is a much more difficult task. By limiting the discussion to interactions between dummy variables and interactions between dummy variables and interval variables, the books keeps the explanation at a concrete level and avoids being too abstract. The book introduces the concept of the within-group model. The book shows that for every standard interaction model, there is a within-group interaction model that the standard model addresses. The book gives the researcher the tools to understand more fully what happens when interaction variables are added to a regression model.

Modeling linearity with splines: Among social scientists, spline variables receive the most attention from economists. My book brings spline variables into the broader discourse in the social sciences and uses spline variables as a way to examine linearity in regression analysis more simply. The common way to model linearity in regression is to introduce a squared variable into the model. A key drawback of this approach is that the coefficient for a squared variable is difficult to interpret and often requires graphing the results. My book suggests using spline variables as an alternative way to examine linearity. The book presents a thorough discussion of two related types of spline models, segment splines and difference splines. The book shows how the results from spline model regressions are immediately interpretable. Thus, the book provides researchers with an alternative and more understandable way to examine linearity than the more commonly used method of squared variables.

Research hypotheses: The last chapter of the book, while obviously at the end, is still an important chapter. The chapter discusses ways to formulate research hypotheses that are amenable to testing by using regression models. The topic of formulating research hypotheses is a topic that receives little attention in most regression books. Research hypotheses connect the literature review to the regression analysis. The material on research hypotheses gives beginning researchers, in particular, a conceptual tool for understanding how a regression model will relate to theoretical issues.

Special Features

The book provides a discussion of regression analysis that takes a different approach from most other books in the following ways:

Analytical tables: Throughout the book, every time an issue of data analysis is discussed, the book presents a table that is directly related. The computer output from a regression analysis is just a first step in a regression analysis. Producing a table that presents the results in an understandable way and then discussing that table is the key step in a regression analysis. The book provides many tables, and most of the tables are like those that would be found in research writing. Thus, the book illustrates throughout what is the end goal of a regression analysis, which is a table and a discussion.

Representing variables as matrices: In the chapters on interaction modeling and control modeling, the book illustrates variables by using a matrix approach. Being able to see how nestedness works in the discussion of within-group and standard interaction models and in the discussion of segment spline and difference spline models is the key to understanding how the various models work. Presenting the variables by using matrices provides the opportunity to visualize the data on which the regression models are based.

Data and statistical code available to replicate models: Reading about regression modeling takes a researcher part of the way to understanding regression modeling. Actually running regression models takes the researcher the rest of the way. The book provides a link to the High School Longitudinal Study data used throughout the book and provides the statistical program code for creating the exact data file used in the book. The book also provides the statistical program code to create all the variables used in the book. The data file and statistical program code are also available on the SAGE website at **http://study .sagepub.com/wojtkiewicz**. Providing these materials gives the researcher the opportunity to read about a statistical modeling approach and then to learn how to run those same models to produce correct results.

Answers to chapter exercises: The chapter exercises ask users of the book to replicate analyses in the book and to create new analyses. The section of the book on answers to chapter exercises presents the results for the new analyses. The opportunity to replicate results and to create new results is the key to fully understanding how to do regression modeling.

So time to get started, good luck, and enjoy your search for results!

Acknowledgments

I wish to thank Leah Fargotstein, Kassie Graves, Yvonne McDuffee, Kelly DeRosa, Sheree Van Vreede, and Vicki Knight at SAGE for their excellent help on the project from beginning to end. I appreciate my professors in the sociology graduate program at the University of Wisconsin–Madison who taught me about data analysis. In particular, I am grateful to Hal Winsborough, my dissertation advisor, and to Jim Sweet for their constant support throughout my career. I thank my students and colleagues at Louisiana State University (LSU) for their support and contributions as I developed the ideas for this book. I am indebted to Wayne Villemez, Joachim Singelmann, and Andy Deseran at LSU for giving me the opportunity to start an academic career. My students and colleagues at Ball State have been instrumental in providing me with an environment where I could work out my ideas on regression analysis. I appreciated very much the support of Jun Xu and Rachel Kraus as I developed my book. I thank Jun Xu, Yanyi Djamba, Dan Powers, and the reviewers for reading the manuscript and for making helpful suggestions. Those reviewers are as follows:

Naveen Bansal, Marquette University

Cynthia M. Cready, University of North Texas

Robert J. Eger III, Naval Postgraduate School, Graduate School of Business and Public Policy

Craig Parks, Washington State University

Guogen Shan, School of Community Health Sciences, University of Nevada Las Vegas

Christine Tartaro, Richard Stockton College

Finally, I am grateful to Elaine, Will, my family, and my friends for making my life a richer one.

About the Author

Roger A. Wojtkiewicz is a professor in the Department of Sociology at Ball State University in Muncie, Indiana. He spent the first 12 years of his career in the Department of Sociology at Louisiana State University (LSU) and has since been at Ball State, where he served as department chairperson for 12 years. At LSU, he taught undergraduate statistics and a graduate course in regression modeling in the PhD program. It was at LSU where he began the development of the ideas contained in this book. At Ball State, he has taught both the first and second semester courses in the statistics sequence in the master's program, and it is here that he finalized the ideas contained in this book. He also uses some of the basic ideas about regression analysis found in this book in the undergraduate statistics course that he teaches each semester. He was trained as a quantitative methodologist in the graduate sociology program at the University of Wisconsin–Madison. His research approach is quantitative, and he uses regression analysis to examine issues in the areas of family demography and educational stratification.

Introductory Ideas

Regression Modeling

An important part of social science work is empirical verification of theoretical ideas. One approach to empirical verification is quantitative research in which numbers are used to express observed aspects of reality. The most important tool for quantitative research in the social sciences is regression analysis. Regression analysis involves using independent variables to explain the mean level of a dependent variable by estimating regression coefficients for the independent variables.

One application of regression analysis is to use sample data to estimate for the population the influence of an independent variable on a dependent variable. However, social scientists use regression analysis for broader purposes than simply producing coefficients to estimate effects of variables. In this book, I examine four broader uses of regression analysis:

- Control modeling
- Modeling interactions
- Using spline variables
- Testing research hypotheses

Each of the broader uses of regression analysis involves an in-depth examination of a regression effect as represented by a single coefficient. Thus, I define regression modeling in this book as the use of one or more regression equations to examine a particular coefficient in greater depth.

Control Modeling

Control modeling involves first estimating a regression equation with particular emphasis on a coefficient for one of the independent variables in the analysis. I refer to this independent variable as the "independent variable of interest." The next step is to estimate one or more regression equations with additional independent variables called "control variables."

Adding the control variables ensures that the effect of the independent variable of interest does not capture the correlated influences of the control variables. Additional models may be estimated in an attempt to "explain" the initial coefficient of the independent variable of interest.

The need for control modeling arises when we analyze secondary data such as data from nationally representative surveys. When analyzing group differences, membership in one group may be related to membership in another group. Control modeling narrows the meaning of the coefficient for the independent variable of interest by removing related influences.

For example, if the regression analysis shows that children in stepparent and single-parent families score lower in academic achievement tests, the next step in the analysis is to control for family income. Family income would be controlled since it is related to both family structure and achievement.

Modeling Interactions

When there are two independent variables in a model without any interaction term, then the effect of one independent variable does not depend on the value of the other independent variable. An additive regression equation has no interactions, and the effects for any particular independent variable are specified to be the same for any subgroup that one might define.

However, there are instances when the researcher hypothesizes that the effect of one independent variable depends on the value of a second independent variable. An example of an interaction is a hypothesis that says that the positive effect of attending private high school on chances of college attendance is greater for lower income students than for higher income students. The effect of attending private school is hypothesized to be different in the two groups.

Testing for interactions by using regression modeling involves first estimating an equation with additive independent variables and then estimating an equation with an interaction between the independent variables. There are different ways to specify interaction models, and I discuss the various ways in this book. One type of interaction model estimates the effects of an independent variable in different subgroups.

A second type of interaction model estimates the differences between subgroups in the effects of an independent variable.

Modeling Linearity With Splines

The first regression model that students usually learn about is one with an interval-level independent variable and an interval-dependent variable. In this book, we carefully consider models like this that involve only interval variables.

However, from the very start in this book, we consider the key assumption in a model that includes an interval independent variable and an interval dependent variable. This assumption is that the relationship between the variables is linear. By linear we mean that as the independent variable increases by one additional unit, the dependent variable increases/decreases by a consistent amount.

Another way to think about the linearity assumption is to view the assumption as a constraint on the regression model. This implies that there is a less constrained model than the linear model and that the less constrained model lessens or perhaps does not involve a linear assumption. In this book, we use spline variables to model linearity. Spline variables involve basically taking the interval variable and breaking it into parts. We can then consider the relationship of each part to the dependent variable. It may be that each part has the same linear relationship with the dependent variable. However, it may not be that each part has the same linear relationship. Spline variables provide an accessible method for assessing whether a linear relationship is present.

The relationship between an interval variable and a dependent variable that measures the log odds of being in one category as opposed to another can also be expressed as a linear coefficient. In this book, I also use spline variables to examine linearity in logistic regression models.

Testing Research Hypotheses

Researchers have a purpose when they estimate regression equations and that purpose is to test research hypotheses. In the social sciences, these hypotheses are usually stated in words. Statistics are then used to test the hypotheses.

Social scientists accept the general scientific idea that hypotheses are not ever proven to be true but at best can only be supported. The argument is that it is always possible that evidence might arise that would contradict a supposedly proven hypothesis. The possibility of contradictory information arising is particularly strong in survey research where research results found with one sample from a population could be contradicted by results from another sample from the same population.

A researcher can use regression analysis to test research hypotheses in several ways, and four of those ways are examined in detail in this book. Research hypotheses can be tested by estimating individual coefficients and by estimating a series of regression equations as part of control modeling, modeling interactions, or in using spline variables to model linearity.

Thus, this book has two main goals. One is to provide an intuitive explanation of regression analysis. The other is to provide thorough guidance on how to use regression analysis to address social science research hypotheses.

Classical Approach to Regression

The relation between an independent variable and a dependent variable using regression analysis is classically depicted as a line drawn through a scatterplot of points.[1] In many disciplines, regression analysis is used primarily to examine the relationship between continuous/interval variables.

A problem for students in the social sciences with this approach to understanding regression is that it does not relate to what the students first learn in the typical statistics course. Social science statistics courses usually start with concrete basic statistics, not with abstract statistics like regression.

Disadvantages of Classical Approach

Statistics courses in the social sciences most often begin with discussions of frequency distributions, cross-tabulations, central tendency, and variation. After some basics of probability and the normal distribution, students then are taught about the difference between means test (also known as the *t* test) and the chi-square test.

Regression analysis is usually introduced after this material is covered by discussing scattergrams and fitted lines. Although the preliminary material in the course involved comparing groups, the scattergram approach involves the relationship between two individual characteristics. The idea of comparing groups is absent. The disadvantage with using the scattergram approach is more than just statistical. A key part of social science thinking is thinking about how groups are different from one another.[2] The scattergram approach is not about groups but is about the linear relationship between different factors.

The discrete approach to regression modeling that I use in this book takes a different approach than the scattergram approach and builds on the group concept in explaining regression analysis. My approach first discusses dummy variable regression that connects directly to the difference between means test. The discrete approach thus builds directly on student knowledge about cross-tabulations and differences between means that is developed early in their statistics courses.

---

1. For an introductory discussion to the classical regression approach, see Gordon (2012, Chap. 5), Kahane (2001, Chap. 2), and Wooldridge (2013, Chap. 2).

2. Agresti and Finlay (2009), Chapter 7, present a basic discussion of group comparison as a basic analytical approach in statistical analysis.

Discrete Approach to Regression

The discrete approach to explaining regression for social scientists begins with the idea that the central question in regression analysis is about how social groups are different from one another. The groups are compared on a dependent variable. Discrete variables are simply variables, whether categorical or interval, that have a finite number of values rather than the theoretically infinite, or at least very many values, as is the case for continuous interval variables.[3]

In this book, I focus on two ways of measuring the dependent variable: the mean and the log odds. Although the groups we often compare in social science research can be defined by categorical variables, the characteristics of groups are often described by both interval and categorical variables. The mean is used to measure interval variables, and the log odds is used to measure categorical variables. We compare groups by taking differences in means or log odds between groups.

Dummy variable regression is a method for analyzing group differences. Although the classical approach to discussing regression analysis starts with interval variables and eventually discusses dummy variables as a special case, the approach in this book starts with dummy variables and then addresses continuous variables as an equal partner. The discrete approach provides the basis for the discussion of control models, interactions, spline variables, and research questions.

Summary

Regression modeling involves the use of one or more regression equations to examine a particular coefficient in greater depth. Control modeling involves adding additional independent variables to a regression equation and examining how the coefficient for the independent variable of interest changes with the addition of the variables. Interaction modeling involves considering how the effect of one independent variable depends or is conditioned on the effect of a second independent variable. Researchers use spline variables to model possible nonlinear effects of an independent variable on a dependent variable.

The reason for using regression modeling techniques is to test research hypotheses. In this book, I use a discrete approach that focuses on group differences to explain regression modeling. The key objective of this book is to provide regression modeling tools that social science researchers can use to test research hypotheses.[4]

---

3. Demaris (2004), page 18, defines a discrete variable as "one with a countable number of values."

4. The focus of this book is on regression modeling. An important concern when using regression is addressing the assumptions of the regression model. Discussion of regression assumptions does not fit into the focus of the present book. See Fox (2015) for an extensive discussion of issues related to regression assumptions.

Key Concepts

regression modeling: the use of one or more regression equations to examine a particular coefficient in greater depth.

control modeling: adding control variables to a regression equation to ensure that the effect of the independent variable of interest does not capture the correlated influences of the control variables.

modeling interactions: using interaction variables to examine whether the effect of one independent variable depends on the value of a second independent variable.

modeling linearity with splines: using spline variables that are created by breaking an interval variable into parts to consider whether the parts are linearly related to the dependent variable each in the same manner.

testing research hypotheses: estimating individual coefficients, estimating a series of regression equations as part of control modeling, modeling interactions, or using spline variables to model linearity to test research hypotheses.

discrete approach to regression: begins with the idea that the central question in regression analysis is about how social groups are different from one another and compares groups on a dependent variable.

2

Basic Statistical Procedures

Individual Units and Groups

In the discrete approach to regression analysis that I use in this book, individual units and groups of individual units have clearly defined roles. The key analytical unit in the discrete approach to regression analysis is the group. Many of the hypotheses that we test using regression analysis in the social sciences usually involve how one group is different than another group in some way. The discrete approach builds directly on the analysis of group differences as in the *t* test or cross-tabulations.

The role of the individual unit of analysis is that individual characteristics are summarized statistically and the summary statistic is used to describe the group.[1] The

---

1. The brief discussion of basic statistics in this book is presented to provide a base for the more complex material that follows. For more thorough discussions of basic statistics, see Frankfort-Nachmias and Leon-Guerrero (2014) and Linneman (2014) at the undergraduate level and Gordon (2012) and Agresti and Finlay (2009) for noncalculus-based discussions at the graduate level.

groups analyzed in regression analysis have no measured characteristics other than the summarized characteristics of the units in the groups.

For example, the mathematics ability of the group defined as students in private schools may be measured by the mean scores of students on a mathematics test. We do not measure directly the mathematical ability of the group but instead summarize the mathematics ability of students in the group and use that statistic to describe the group. The social science question addressed by statistical analysis is not whether individuals are different from one another but whether the groups that individuals belong to are different from one another.

Measurement

Measurement involves assigning to a unit of analysis a value that captures a certain characteristic of a unit of analysis. For example, the number of credit hours taken in a semester may be assigned to each student, and then the mean number of credit hours taken by students can be calculated. Likewise, each student can be assigned to one of the categories of type of major, and then the percentage of students in each type of major can be calculated.

Researchers refer to the characteristic of the units of analysis that is being measured as a variable. Thus, a variable is a characteristic of a unit that will vary across units. The word *variable* implies that a characteristic varies across units of analysis. Social scientists generally have no interest in analyzing characteristics that do not vary across units of analysis. Variables contain different amounts of information. Some are merely names, others are names that have hierarchical order, and still others provide measurements in standard units.

Collecting data involves the measurement of variables. One widely used method for collecting data is survey research. Survey research is where a respondent provides information that allows measurement of a variable about a unit of analysis. The unit of analysis may be the respondent or some other unit. The examples in this book use public-use data from the High School Longitudinal Study of 2009 (HSLS). In the data used here, the basic unit of analysis is a student who was a junior in high school in the United States in the spring of 2012. The student provided information for variables as did parents.

Level of Measurement

Measures themselves have characteristics, and the characteristics of measures relevant to statistical analysis are called the level of measurement of a variable. These characteristics can be framed as questions: Does the variable categorize units, does the variable order units, and does the variable capture a meaningful numerical amount of a characteristic that a unit possesses?

A variable that only categorizes is at the nominal level, a variable that categorizes and orders is at the ordinal level, and a variable that categorizes, orders, and captures a meaningful numerical amount is at the interval level.

	Categorize	Order	Numerical
Nominal	Yes	No	No
Ordinal	Yes	Yes	No
Interval	Yes	Yes	Yes

The nominal level of measurement provides the least amount of information since it does not allow us to rank groups or compare groups based on how much of a characteristic they might have. Ordinal adds more information in allowing for the ordering of categories. Interval adds more information by allowing for the amount of the characteristic to be considered. Be aware that the ratio level is a fourth level of measurement. The ratio level adds a true zero to the interval level, but having a true zero is not a major issue in the analysis of survey data using regression procedures since the focus is on differences in levels and not on the levels themselves. In this book, I use the term *interval* to refer to both interval- and ratio-level variables.

Examples for Level of Measurement

The nominal level categorizes without order or meaningful numerical value:

School control:
> public, Catholic, other private

Sex/gender:
> female or male

Family structure:
> two biological parents, biological/stepparent, single parent, other family

The ordinal level categorizes with order but without meaningful numerical value:

Socioeconomic status (SES) quartile:
> lowest, second, third, highest

Parental education:
> < high school, high school, BA/BS, master's degree, PhD or higher

The interval level provides meaningful numerical values:

Standardized score—math:
> 20.91, . . . , 81.04

Number of siblings:
> 0, 1, 2, 3, 4, 5, 6, . . .

Count, Sum, and Transformations

The population is the complete set of units of analysis. These units are the object of our study. A sample is a subset of the population. For the sample to be representative of the population without bias, the sample needs to be drawn by using some form of random selection.

The frequency is simply the number of units of analysis in the sample, in a category of a variable, or of a particular value. Frequencies are used to describe nominal and ordinal variables and are used to describe interval variables when the number of values of the interval variable is not too large. A frequency is simply a count. How many are there in total in the sample? How many in the sample attend private schools? How many in the sample have exactly three siblings? By construction, the sum of the frequencies for each value of a variable will add to the total.

The sum is used for interval variables and is the sum of values of the interval variable in the sample or in subgroups of the sample. Often the subgroups for which the sum is calculated are categories of a nominal or ordinal variable. We could take the sum of the number of siblings for the whole sample, or we could take the sum for those in public schools.

Statistical procedures often involve mathematical transformations. Linear transformations involve multiplying, dividing, adding, and subtracting. Nonlinear transformations include squaring, square root, and natural logarithm.

Mean

The mean is the sum of the values on an interval variable divided by the number of values. By dividing the sum by the number of units of analysis, the mean produces a number that expresses the amount of the variable per unit of analysis:

$$\bar{X} = \Sigma \frac{X}{N}$$

where

$\bar{X}$ is the sample mean

ΣX is the sum of all Xs in the sample

N is the number of units in the sample

The mean is a redistributive statistic. The mean can be viewed as adding all the X values up and then giving the same amount of X to each sample unit.

The mean combines two statistics: A sum is divided by a frequency. We typically do not use only the sum to describe a sample because the sum can be influenced by the size of the sample. The mean takes sample size into account by dividing the sum by N, the frequency of units in the sample.

A statistic that can be used with interval and ordinal variables is the median. The median is a value that divides a sample into two equal parts in terms of frequency. The median provides information about the sample that is easy to understand, perhaps more so than the mean. However, the mean is used in advanced statistics because it is amenable to mathematical manipulation and the mean is the minimum variance measure of central tendency. The median, on the other hand, does not have these properties.

Proportion and Percentage

The proportion and percentage are two closely related statistics that we can use to describe a nominal or interval variable. The constraint on using proportions and percentages to describe interval variables is that the number of values for the interval variable needs to be reasonably small. At the extreme, if each value of the interval variable is unique, there will be as many proportions or percentages as there are values. The resulting proportions or percentages will be too many in number to be useful for analytical purposes:

$$P = \frac{f}{N}$$

$$\% = (\frac{f}{N}) \times 100$$

where

P is the proportion

f is the frequency in a category

N is the number of units in the sample

$\%$ is the percentage

A simple statistical procedure is the frequency distribution. The frequency distribution provides the count for each category of a variable. Since sample sizes can vary, the frequency in a particular category is difficult to assess in terms of relative size. The proportion divides f by N and provides a statistic that says if the total N is 1.00, how much of the 1.00 is in each category. We calculate the percentage by multiplying the proportion by 100. The percentage provides a statistic that says if the total N is 100, how much of the 100 is in each category. The proportion and the percentage provide essentially the same information, differing only in units expressed. However, the percentage is the more widely used descriptive statistic.

Odds and Log Odds

Linear regression, also known as ordinary least-squares regression, is appropriate when the dependent variable is interval. It is possible to apply ordinary least-squares regression to a nominal or ordinal variable coded 0 or 1 (called a dichotomous variable). However, there are problems statistically in using linear regression with a dichotomous dependent variable. Among these problems is that the predicted values when using linear regression with a dichotomous dependent variable may fall outside the range of possible values such as predicting a negative proportion or a proportion more than 1.0. Another problem is that the standard errors that linear regression will estimate for the regression coefficients are biased since they do not on average capture the population standard error.[2]

Linear regression with an interval dependent variable can be viewed as a method for comparing means. Research life would be simpler if we could use linear regression with a dichotomous dependent variable as a method for comparing proportions. Since using linear regression with a dichotomous variable has major drawbacks, statisticians have devised an approach for analyzing dichotomous variables called logistic regression that alleviates these drawbacks.

We can view logistic regression as a method for comparing log odds. Using log odds in logistic regression makes logistic regression more difficult to understand because we do not typically use log odds for analyzing frequency distributions.

Although the proportion compares f with the total N in a sample or a subgroup of a sample, the odds ratio compares f with "not f." That is, if f is the frequency in a group, not f is the frequency not in the group or is $N - f$:

$$\text{Odds} = \frac{f_1}{f_2}$$

where

f_1 is frequency in group

f_2 is frequency not in group

$f_1 + f_2 = N$

The odds ratio in horse racing describes a horse's chances of losing. An odds of 4:1 indicates that a horse is expected to lose four times for each win.

However, the dependent variable in logistic regression is not the odds, which would be difficult enough to interpret, but the dependent variable is the log odds, where the log is taken to the base e and is thus the natural logarithm (ln):

$$\text{Log odds} = \ln\left(\frac{f_1}{f_2}\right)$$

---

2. See Fox (2015), Chapter 14, and Kutner, Nachtsheim, Neter, and Li (2005), Chapter 14, for fuller explanations about the necessity for using logistic regression when the dependent variable is dichotomous.

where

ln is the natural logarithm to the base e

Logistic regression is the standard regression method when the dependent variable is dichotomous. Logistic regression provides predicted values within the range of possible values and unbiased standard errors for the regression coefficient can be calculated.

I emphasize linear regression and logistic regression in this book. Understanding these two procedures provides a solid base for further use of regression analysis.

Table 2.1	Mean Math Score by Type of School	
Type of School	**Mean Math Score**	**N**
Public	50.80	16,797
Private	55.86	3,336
Total	51.63	20,133

Examples of Means and Log Odds

The respondents in the High School Longitudinal Study took achievement tests for mathematics when they were juniors in high school in 2012 as part of the first follow-up of respondents in the study. The math score for each student was standardized to a national mean of 50. The result was a mean standardized math score of 50 for the total sample.

The mean for each subgroup is calculated by summing scores in each subgroup and dividing that sum by the subgroup total. The data show that students attending private school scored higher in math than students in public schools, 55.86 compared with 50.80. Note that the mean for the total is not 50. This occurs because some members of the original sample were not included in the analytical sample as a result of missing data. Those respondents who had scores less than 50 were overrepresented among the respondents who were excluded from the working sample as a result of missing data, and this led to a sample mean of greater than 50.

Table 2.2 shows a frequency cross-tabulation for those in public and private schools for whether the respondent was in the top 25% on math score. Thus, the top 25% variable is a dichotomous or two-category variable that involves dividing the sample respondents into the top 25% or not top 25% in math score.

The proportion converts frequencies into numbers that are relative rather than absolute. The proportion in the top 25% on math score is higher for those in private

Table 2.2	Frequencies for Top 25% in Math by Type of School		

	Math Score		
Type of School	Top 25%	Not Top 25%	Total
Public	3,730	13,067	16,797
Private	1,295	2,041	3,336
Total	5,025	15,108	20,133

Table 2.3	Proportions for Top 25% in Math by Type of School		

	Math Score		
Type of School	Top 25%	Not Top 25%	Total
Public	.22	.78	1.00
Private	.39	.61	1.00
Total	.25	.75	1.00

Table 2.4	Percentages for Top 25% in Math by Type of School		

	Math Score		
Type of School	Top 25%	Not Top 25%	Total
Public	22	78	100
Private	39	61	100
Total	25	75	100

schools compared with those in public schools, a .39 proportion for private compared with .22 for public.

For those in private schools, 39% were in the top 25% on math score compared with 22% for those in public schools. Those in private are more likely to be in the top 25% than those in public.

The odds calculation for public was 3,730/13,067 = .28. The odds compares a frequency f with its complement $N - f$. Odds less than 1.00 indicate the f is less than

Table 2.5	Odds for Top 25% in Math by Type of School		

	Math Score		
Type of School	**Top 25%**	**Not Top 25%**	**Odds**
Public	3,730	13,067	.28
Private	1,295	2,041	.63
Total	5,025	15,108	.33

$N - f$, whereas odds greater than 1.00 indicate that f is greater than $N - f$. For those in private schools, the odds of being in the top 25% on math score was .63 compared with an odds of .28 for those in public schools. This result also indicates that students in private school are more likely to be in the top 25% than those in public.

The log odds calculation for those in public school is $\ln(3,730/13,067) = -1.25$. The log odds compares a frequency f with $N - f$ in logarithmic terms to the base e. Log odds less than .00 indicates that f is less than $N - f$, whereas log odds greater than .00 indicates that f is greater than $N - f$. A log odds of .00 indicates that f equals $N - f$. One advantage of the log odds compared with the proportion and the odds is that the log odds does not have a predefined lower limit or upper limit. The proportion ranges from 0.0 to 1.0, and the odds ranges from 0.0 to infinity. The log odds ranges from infinitely negative to infinitely positive. Note that the $\ln(0.0)$ is not defined mathematically.

The log odds will show the same relationship between two groups as the percentages. The percentage in the top 25% on math score is higher for those in private school compared with those in public school, 39 compared with 22. Likewise, those in private school had a higher log odds of being in the top 25% on math score than those in public school, $-.45$ compared with -1.25.

Table 2.6	Log Odds for Top 25% in Math by Type of School		

	Math Score		
Type of School	**Top 25%**	**Not Top 25%**	**Log Odds**
Public	3,730	13,067	-1.25
Private	1,295	2,041	$-.45$
Total	5,025	15,108	-1.10

Differences

Table 2.7	Mean Math Score by Type of School	
Type of School	**Mean Math Score**	
Public	50.80	
Private	55.86	

Taking a difference between two means involves subtracting one mean from the other mean. The difference shows how much one mean is more or less than the other mean. The order of subtraction matters for interpretation. The sign of the difference will reverse if the order of subtraction is reversed. Of course, if the magnitude of the difference is measured by absolute value, then the difference will not change when the order of subtraction is changed:

$$\text{Difference} = \bar{X}_1 - \bar{X}_2$$

When only two means are involved in the analysis, reversing the order of subtraction simply reverses the sign of the difference. The mean for those in private schools is 5.06 and is larger than the mean for those in public schools. Those respondents in private schools have a higher mean math score than those in public schools, or those in public schools have a lower mean math score than those in private schools.

We usually analyze a cross-tabulation by using percentages rather than log odds. However, as preparation for understanding logistic regression, I analyze the cross-tabulation in Table 2.9 by using log odds. Those students in private school are more likely to be in the top 25% on math score than those in public schools, $-.45$ compared with -1.25:

$$\text{Difference} = \log \text{odds}_1 - \log \text{odds}_2$$

Table 2.8	Differences in Mean Math Score by Type of School	
$\bar{X}_1$	$\bar{X}_2$	
	Public	**Private**
Public	.00	−5.06
Private	5.06	.00

Table 2.9	Log Odds for Top 25% in Math by Type of School

Type of School	Log Odds Top 25% Math
Public	−1.25
Private	−.45

Table 2.10	Differences in Log Odds for Top 25% in Math by Type of School

	Log Odds$_2$	
Log Odds$_1$	Public	Private
Public	.00	−.80
Private	.80	.00

The difference in log odds is .80 for the private school log odds minus the public school log odds and −.80 for public minus private. Reversing the order of subtraction only results in a reversal of the sign of the difference between log odds. The order of subtraction does not matter much in understanding the information provided when comparing two groups.

The challenge in analyzing differences in means for the four family structure categories shown in Table 2.11 is to decide which differences to focus on. Changing the contrast group changes the differences:

$$\text{Difference} = \bar{X}_1 - \bar{X}_2$$

Table 2.11	Mean Math Score by Family Structure

Family Structure	Mean Math Score
Two Bio.	52.88
Bio./Step.	50.32
Single	50.14
Other Fam.	49.10

Table 2.12	Differences in Mean Math Score by Family Structure			
$\bar{X}_1$	$\bar{X}_2$			
	Two Bio.	**Bio./Step.**	**Single**	**Other**
Two Bio.	.00	2.56	2.74	3.78
Bio./Step.	−2.56	.00	.18	1.22
Single	−2.74	−.18	.00	1.04
Other	−3.78	−1.22	−1.04	.00

With four categories in the family structure variable, there are 12 possible differences but only 6 unique ones. In each column, the mean for a different group is subtracted. Each column represents a different set of differences. The mean differences in the top half of the matrix reflect the bottom half but with a change in sign.

Table 2.13	Log Odds for Top 25% in Math by Family Structure
Family Structure	**Log Odds Top 25% Math**
Two Bio.	−.90
Bio./Step.	−1.35
Single	−1.38
Other Fam.	−1.59

Since only one set of differences is usually considered, a key decision in the analysis of differences is which mean or log odds will be the one that is subtracted. Standard practice is to use mean or log odds for the most analytically important group when subtracting:

$$\text{Difference} = \log \text{odds}_1 - \log \text{odds}_2$$

Since the differences are calculated as the difference from some particular group, the group chosen as the one subtracted is a key decision. All the differences reflect how the means or log odds compare with that one contrast group.

Table 2.14	**Differences in Log Odds for Top 25% in Math by Family Structure**			

	Log Odds₂			
Log Odds₁	**Two Bio.**	**Bio./Step.**	**Single**	**Other**
Two Bio.	.00	.45	.48	.69
Bio./Step.	−.45	.00	.03	.24
Single	−.48	−.03	.00	.21
Other	−.69	−.24	−.21	.00

Summary

Although the unit of analysis in quantitative social science research is often an individual, the analytical unit is a group of individuals. The basic analytical approach in social science research is to describe differences between groups. Measurement involves assigning a characteristic or a value of a variable to a unit of analysis. Measurements can range from simply names as in the nominal level, to ordered names as in the ordinal level, and then to meaningful numerical values as in the interval level.

Researchers can count the number of particular values for nominal, ordinal, and interval variables in a group, while also summing up interval variables. The mean compares a sum of a variable relative to the sample size N, whereas proportions and percentages compare counts of a particular value of a variable relative to sample size N. Alternative measures involving counts are the odds that compares the count for one value of a variable with the count for a second value of a variable. The log odds is the natural logarithm of the odds. Researchers compare groups by taking the difference in means or log odds. Analyzing differences is a key task in social science research.

Key Concepts

units and groups: the basic unit for purposes of measurement is the unit of analysis; a group is a set of units of analysis defined by a common characteristic such as a value of a variable.

measurement: the process of assigning a value of a variable to a unit of analysis.

level of measurement: the values of a variable can be names that are nominal-level values, ordered names that are ordinal-level values, or meaningful numbers that are interval-level values.

count: the number of units in a group with a particular value of a variable.

sum: the additive total of the values of an interval variable in a group.

mean: the sum of the values of a particular variable in a group divided by how many units of analysis are in the group N.

proportion/percentage: the count of units with a particular value of a variable in a group divided by the number of units of analysis in the group N.

odds: the count of units with a particular value of a variable in a group divided by the count of units with another particular value of a variable in the group.

log odds: the natural logarithm of an odds.

difference: the mean, proportion, or log odds for one group minus the mean, proportion, or log odds for a second group.

Chapter Exercises

1. Appendix A describes how to access the High School Longitudinal Study data that are used to create the examples in this book. Access the data at the National Center for Educational Statistics website and download the variables as specified. The data are also available on the SAGE website for this book.

2. Create the working dataset, as specified in Appendix A, by selecting only those cases where W2STUDENT>0 and X2CONTROL>-6. The resulting dataset should include 20,133 cases.

3. Re-create the table of differences where the independent variable is family structure (FAMSTRUCT) and the dependent variable is math score (X2TXMTSCOR).

4. Re-create the table of differences where the independent variable is family structure and the dependent variable is the log-odds of top 25% in math (HIGHMATH).

5. Create a new table of differences where the independent variable is family structure and the dependent variable is socioeconomic status (X2SES). How do students vary in SES by family structure category?

6. Create a new table of differences where the independent variable is family structure and the dependent variable is private high school (PRIVATE). How do students vary in the log-odds of attending private high school by family structure category?

Regression Modeling Basics

Difference Between Means: The *t* Test

One of the most widely used statistical tests in social science research is the difference between means test or, as it is commonly called, the *t* test. The top of the *t* test statistic is simply the difference between means. The bottom is an estimate of the standard deviation in the difference between means distribution. The estimate assumes equal population variances and involves combining the sums of squares in each sample and using the *N*s from each sample to produce the estimate:

$$t = \frac{\overline{X}_1 - \overline{X}_2}{s_{\overline{X}_1 - \overline{X}_2}}$$

$$s^2_{\text{pooled}} = \frac{(N_1 - 1)s_1^2 + (N_2 - 1)s_2^2}{N_1 + N_2 - 2}$$

$$s_{\bar{X}_1 - \bar{X}_2} = \sqrt{\frac{s^2_{pooled}}{N_1} + \frac{s^2_{pooled}}{N_2}}$$

The t test compares the actual difference between sample means with an estimate of the sample variation in that difference. A large t test statistic will occur when the difference between means is large relative to expected sample variation in that difference.

A more common computational formula for $s_{\bar{X}_1 - \bar{X}_2}$, algebraically equal to the definitional formula, is as follows:

$$s_{\bar{X}_1 - \bar{X}_2} = \sqrt{\left(\frac{(N_1 - 1)s_1^2 + (N_2 - 1)s_2^2}{N_1 + N_2 - 2}\right)\left(\frac{N_1 + N_2}{N_1 N_2}\right)}$$

Table 3.1 Mean Math Score by Type of School			
Type of School	**Mean Math Score**	**N**	**SE**
Public	50.80	16,797	10.12
Private	55.86	3,336	8.89

The following is the computation for the t statistic that uses the definitional formula. The formula assumes equal population variances that is the same assumption as in linear regression:

$$s^2_{pooled} = \frac{(16,797 - 1)(10.12^2) + (3,336 - 1)(8.89^2)}{16,797 + 3,336 - 2} = 98.54$$

$$s_{\bar{X}_1 - \bar{X}_2} = \sqrt{\frac{98.54}{16,797} + \frac{98.54}{3,336}} = .19$$

$$t = \frac{55.86 - 50.80}{.19} = \frac{5.06}{.19} = 26.63$$

The critical t value required for a decision that the difference between means in the population is not equal to zero at the .05 level is 1.96. Since 1.96 is about equal to 2, the sample difference must be about twice as large as the standard error for the sample difference to reject at the .05 level the hypothesis that the difference in the population is zero.

In this case, the t value is 26.63, so the difference is more than 25 times as large as the standard error. The result of this test is that the sample difference is large enough to say that the population difference is likely not equal to zero.

Linear Regression With a Two-Category Independent Variable

Table 3.2 provides the first regression model that we will discuss. This linear regression includes a standardized math score as the dependent variable and a dummy variable for private school as the only independent variable.[1]

A dummy variable is a variable that is coded "1" if a respondent is in the category of interest and "0" otherwise. In this case, all the respondents who were attending private school are coded as "1" and those attending public school are coded as "0."

Table 3.2 Linear Regression of Type of School on Math Score			
Independent Variable	*B*	**SE**	*t*
Private	5.06	.19	26.63
Intercept	50.80	.08	—

There are familiar numbers in the regression results.[2] The *B* coefficient is 5.06 and is the difference between the mean for those in public school subtracted from the mean for those in private school. The standard error for *B* is .19 and is the same as the standard error in the *t* test. The *t* value in the regression is 26.63 and is the same as the *t* value in the *t* test. These results show that using a single dummy variable in linear regression produces the same results as a *t* test that assumed equal population variances. Notice that the intercept in this regression is 50.80, which is exactly equal to the mean for those in public school. I will say more about the intercept later in this chapter and in Chapter 4.

Table 3.3 provides a similar model as that in Table 3.2 but includes a dummy variable for public school rather than a dummy variable for private school. In this case, all

Table 3.3 Linear Regression of Type of School on Math Score			
Independent Variable	*B*	**SE**	*t*
Public	−5.06	.19	−26.63
Intercept	55.86	.17	—

1. Allison (1999) provides a thorough, introductory discussion of the technical aspects of regression analysis. Kahane (2001) provides a comprehensive discussion of the basics of regression analysis without using calculus or matrix algebra.

2. Linneman (2014), pages 327–328, and Gordon (2010), page 213, also show that a linear regression with one dummy variable produces the same result as the *t* test.

the respondents who were attending public school are coded as "1" and those attending private school are coded as "0."

The *B* coefficient is −5.06 and is the negative of the *B* in the model that included the dummy variable for private school. It makes sense that the sign of the coefficient is reversed since we are just subtracting the private school mean from the public school mean rather than subtracting the public mean from the private mean. The standard error is the same in this model as in the model that included private school. The *t* value is reversed in sign since the difference between the means is reversed in sign.

However, the intercepts are different in the two models. The intercept in the second model is the private school mean, whereas the intercept in the first model was the public school mean. Thus, when using a dummy variable in linear regression, the intercept is the mean for the contrast group.

Table 3.4 shows the means for those in public school and for those in private school. The mean of 50.80 for those in public school is lower than the mean of 55.86 for those in private school. We will see these numbers in the regression results.

Table 3.4 Mean Math Score by Type of School	
Type of School	**Mean Math Score**
Public	50.80
Private	55.86

I have combined the results from two regression models into one table. Again, the intercept in the model with the dummy variable for private school is the mean for public school and the intercept in the model with the dummy variable for public school is the mean for private school. The coefficient in Model 1 is the reverse from the coefficient in Model 2 because the coefficient in Model 1 reflects the private mean minus the public mean, whereas the coefficient in Model 2 reflects the public mean minus the

Table 3.5 Linear Regression of Type of School on Math Score				
	Model			
Independent Variable	**1**		**2**	
	B	**SE**	***B***	**SE**
Private	5.06	.19	—	—
Public	—	—	−5.06	.19
Intercept	50.80	.08	55.86	.17

private mean. The two regression models provide similar information. The coefficient in Model 1 indicates that those in private school score 5.06 higher in math than those in public school. The coefficient in Model 2 indicates that those in public school score 5.06 lower in math than those in private school.

Logistic Regression With a Two-Category Independent Variable

Table 3.6 shows the log odds of being in the top 25% in math for those in public school and those in private school. The log odds are −1.25 for those in public school and −.45 for those in private school. As in the dummy variable linear regression, we will see these log odds in the dummy variable logistic regression results.[3]

Table 3.6	Log Odds for Top 25% in Math by Type of School

Type of School	Log Odds Top 25% Math
Public	−1.25
Private	−.45

Table 3.7 shows the combined results from the two regression models. The intercept in the model using the private variable is the log odds for those in public schools, and the intercept in the model with the public variable is the log odds for those in private schools. The coefficient in Model 1 is the reverse from the coefficient in Model 2

Table 3.7	Logistic Regression of Type of School on Top 25% in Math

Independent Variable	Model			
	1		2	
	B	SE	B	SE
Private	.80	.04	—	—
Public	—	—	−.80	.04
Intercept	−1.25	.02	−.45	.04

3. See Pampel (2000) for a thorough introduction to logistic regression and the related technique, probit analysis. Demaris (1992), pages 42–79, and Jaccard (2001), Chapter 1, provide other useful explanations.

because the coefficient in Model 1 reflects the private log odds minus the public log odds, whereas the coefficient in Model 2 reflects the public log odds minus the private log odds. The two regression models provide similar information. The coefficient in Model 1 indicates that those in private school have a .80 higher log odds of being in the top 25% in math than those in public school. The coefficient in Model 2 indicates that those in public school have a .80 lower log odds of being in the top 25% in math than do those in private school.

Linear Regression With a Four-Category Independent Variable

Table 3.8 shows the means in math test scores by family structure category. We will use linear regression to decide whether the means are significantly different from one another. We need to test for the statistical significance of differences because the size of the means can fluctuate from sample to sample, and thus so will the differences between sample means.

Table 3.9 shows four possible sets of differences between sample means. Each set uses a different mean as the contrast group. Each number in the matrix represents a difference. The numbers above the diagonal are a reflection of those below with only a change in sign.

Table 3.8 Mean Math Score by Family Structure

Family Structure	Mean Math Score
Two Bio.	52.88
Bio./Step.	50.32
Single	50.14
Other Fam.	49.10

Table 3.9 Differences in Mean Math Score by Family Structure

$\overline{X}_1$	$\overline{X}_2$			
	Two Bio.	**Bio./Step.**	**Single**	**Other**
Two Bio.	.00	2.56	2.74	3.78
Bio./Step.	−2.56	.00	.18	1.22
Single	−2.74	−.18	.00	1.04
Other	−3.78	−1.22	−1.04	.00

Table 3.10 shows the results for four linear regressions. Each regression includes three dummy variables for family structure. The category not included in the model is the contrast group.

The results replicate the differences between means discussed earlier. The intercept coefficient captures the mean for the excluded category, whereas the coefficients for the dummy variables capture the difference between the mean for the category represented by the dummy variable and the mean for the excluded category.[4] For example, the coefficient for the dummy variable for single in Model 1 is −2.74, which is equal to 50.14 minus 52.88. The interpretation of dummy variable coefficients in linear regression is straightforward since the coefficient captures the difference between two means. The asterisk (*) indicates statistical significance at the .05 level.

The four models provide varying views on the differences between respondents in the four family structure groups. Model 1 shows that those living with two biological parents score significantly higher than those in the other three groups. Models 2 and 3 indicate that those living with a biological parent/stepparent or with a single parent have equal math scores but have significantly lower math scores than those living with two biological parents and higher math scores than those in other family situations. Model 4 shows that those living in other family situations score significantly lower in math than those in the other three groups.

The choice of which model to estimate depends on which differences are most important for answering the research question. Model 4 is usually not of great interest

Table 3.10	Linear Regression of Family Structure on Math Score

Independent Variable	Model			
	1	2	3	4
	B	*B*	*B*	*B*
Two Bio.	—	2.56*	2.74*	3.78*
Bio./Step.	−2.56*	—	.18	1.22*
Single	−2.74*	−.18	—	1.04*
Other Fam.	−3.78*	−1.22*	−1.04*	—
Intercept	52.88	50.32	50.14	49.10

*$p < .05$.

4. The procedure for testing multiple differences by changing the reference group is illustrated in Gordon (2010), pages 218–220.

since the other family category is included in the analysis to allow the other three categories to be more homogeneous. One would choose Model 1 if the emphasis in the research question is on how those living in two-parent families are advantaged in math scores compared with the other three groups. Models 2 or 3 might be chosen if either the relative position of those living with a biological parent/stepparent or the relative position of those living with a single parent is the key issue.

Logistic Regression With a Four-Category Independent Variable

Table 3.11 shows the log odds for top 25% in math score for each family structure category. Those students in two-biological-parent families have higher log odds of being in the top 25% than those in biological-parent/stepparent families, single-parent families, or other situations. The pattern of difference in the log odds of being in the top 25% in math is similar to the pattern of differences between the groups in mean math scores.

Table 3.12 shows the differences in log odds among the four family structure categories. The difference equals log odds$_1$ minus log odds$_2$. As in the linear regression

Table 3.11   Log Odds for Top 25% in Math by Family Structure

Family Structure	Log Odds Top 25% Math
Two Bio.	−.90
Bio./Step.	−1.35
Single	−1.38
Other Fam.	−1.59

Table 3.12   Differences in Log Odds for Top 25% in Math by Family Structure

Log Odds$_1$	Log Odds$_2$			
	Two Bio.	Bio./Step.	Single	Other
Two Bio.	.00	.45	.48	.69
Bio./Step.	−.45	.00	.03	.24
Single	−.48	−.03	.00	.21
Other	−.69	−.24	−.21	.00

example, there are four sets of differences. As was the case with the mean differences in math scores, there is only a small difference in log odds between the biological-parent/ stepparent and single-parent family categories.

Table 3.13 shows the results for four logistic regressions in which three dummy variables for family structure are included in each model. These models replicate the differences in log odds of being in the top 25% on math score that appear in Table 3.12. The coefficients for dummy variables in logistic regression represent differences like the coefficients for dummy variables in linear regression but represent differences in log odds instead of means.

The family structure category that is not included in the model is the group against which the other family structure groups are compared. This makes that group perhaps the most important group in the analysis since it is the one that all the other groups are compared against. A wise rule for choosing a contrast group in dummy variable regression is to pick a very meaningful group to compare against. The choice will depend on the research question.

Model 1 shows that those living with two biological parents have significantly higher log odds of being in the top 25% in math than those in the other three groups. Models 2 and 3 indicate that those living with a biological parent/stepparent or with a single parent have equal log odds of being in the top 25% in math but have significantly lower log odds than those living with two biological parents and higher log odds than those in other family situations. Model 4 shows that those living in other family situations have significantly lower log odds of being in the top 25% in math than those in the other three groups.

Table 3.13 Logistic Regression of Family Structure on Top 25% in Math

Independent Variable	Model			
	1	2	3	4
	B	*B*	*B*	*B*
Two Bio.	—	.45*	.48*	.69*
Bio./Step.	−.45*	—	.03	.24*
Single	−.48*	−.03	—	.21*
Other Fam.	−.69*	−.24*	−.21*	—
Intercept	−.90	−1.35	−1.38	−1.59

*$p < .05$.

Modeling Linear Effect With Dummy Variables

The effects of interval variables can be modeled in regression analysis by using dummy variables.[5] A dummy variable can represent one value, or a dummy variable can represent groups of values. Table 3.14 shows the means on math score for family socioeconomic status (SES) quintile. SES quintile is a five-category ordinal variable that I treat as an interval variable in this example. Math scores increase steadily as SES increases from quintile 1 to quintile 5.

Table 3.14	Mean Math Score by SES Quintile
SES Quintile	**Mean Math Score**
SES Q1	46.38
SES Q2	48.55
SES Q3	49.99
SES Q4	52.26
SES Q5	57.39

Table 3.15 shows the results from the regression in which dummy variables for second, third, fourth, and fifth family SES quintile were included in the model. The coefficient for the difference increases with each additional quintile.

Table 3.15	Linear Regression of SES Quintile on Math Score		
Independent Variable	**B**	**SE**	**t**
SES Q2	2.17	.23	9.43
SES Q3	3.61	.23	15.70
SES Q4	5.88	.22	26.73
SES Q5	11.01	.21	52.43
Intercept	46.38	.15	—

5. Using dummy variables to model nonlinearity is discussed in Demaris (2004), pages 288–290, and Gordon (2010), pages 321–322.

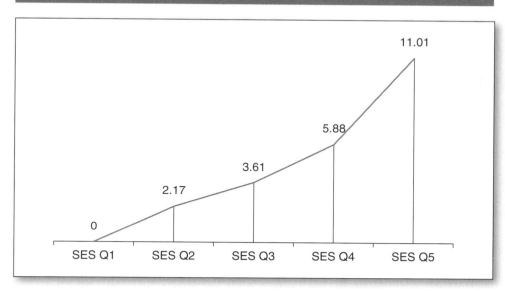

Figure 3.1   Dummy Variable Coefficients for SES Quintile Effects on Math Score

Figure 3.1 shows a graph of the dummy variable coefficients for the effects of SES quintile on math score. The graph has a vertical line drawn to show the size of each coefficient. The vertical lines are then connected by straight lines. In the case of SES quintile, the overall line is nearly straight although the line does veer at a sharper upward angle between Quintile 4 and Quintile 5. By nearly straight, I mean that the line in this case has positive slopes throughout. In regression analysis, the concept of straightness in such a line is called linearity.

In regression analysis, the usual approach to modeling the effect of an interval independent variable is simply to include the interval variable in the model. The result is one coefficent that estimates the effect of each additional unit of the interval variable on the dependent variable. In this example, it is the effect of each additional unit of the SES quintile variable. It will be a linear effect because a line connecting estimated sub-group means would be a straight line.

Linear Coefficient in Linear Regression

Table 3.16 shows the regression results when the interval SES quintile variable is used to predict math score. The linear coefficient in this model is calculated by using the principle of least squares. The linear coefficient shows that each additional unit of the SES quintile variable increases math scores by 2.67 units.

The formula for B in Table 3.16 is as follows:

$$B = \Sigma\left(X - \bar{X}\right)\left(Y - \bar{Y}\right) / \Sigma(X - \bar{X})^2$$

Table 3.16	Linear Regression of SES Quintile on Math Score		
Independent Variable	**B**	**SE**	**t**
SES Quintile	2.67	.05	53.40
Intercept	42.91	.17	—

Statisticians have proved by using the mathematics of calculus that this formula produces a *B* coefficient that when used to create predicted means produces predicted means with the minimum squared deviation from actual means.[6] This is why this formula for *B* is called the "least squares" estimator. The *B* in this case is a linear coefficient that indicates that for every unit increase in the independent variable, there is the same amount of increase in the dependent variable. A line drawn connecting estimates would have a constant slope.

In Figure 3.2, a line representing the linear coefficient is drawn in. In this case, the line connecting the coefficients from the dummy variable analysis is close to the line

Figure 3.2   Dummy Variable Coefficients for SES Quintile Effects on Math Score With Line for Linear Effect

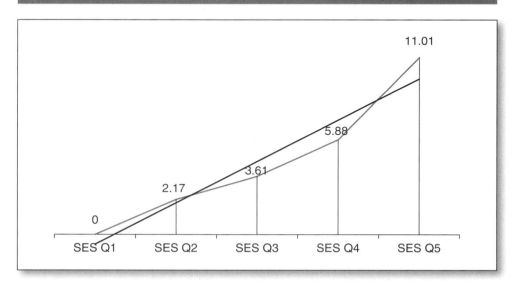

6. Gordon (2010), pages 94–110, presents an accessible discussion of the linear coefficient based on the standard least-squares explanation. Other thorough treatments include Agresti and Finlay (2009) and Fox (2015). Wooldridge (2013) presents an explanation of regression analysis that uses matrix algebra.

representing the linear coefficient. This indicates that the linear coefficient is a good representation of the relationship between SES quintile and math score.

The use of the linear coefficient requires that the variable be measured at the interval level. Sometimes ordinal variables are used in regression as if they are interval variables. In fact, SES quintile is an ordinal variable that I am using as an interval variable. Another example of an ordinal variable that is often used by researchers as if it is an interval variable is parental education, which in the High School Longitudinal Study of 2009 (HSLS) data has eight categories.

Table 3.17 shows the result of a linear regression where seven dummy variables were included to model the effect of parental education on math score. I treat this ordinal variable as if it were an interval variable in this example. The dummy variable for less than high school was not included in the model and serves as the contrast group. The contrast group is the category against which all the other categories are compared. The choice of the contrast group is a key decision when using dummy variables in regression modeling.

The estimated difference between each category in the model and the category for less than high school increases as the level of parental education increases. All coefficients are significantly different from zero. The sample results support the idea that the respective means for parental education categories in the population are not equal to the mean for the less-than-high-school category.

Figure 3.3 shows a graph of the dummy variable coefficients for the effects of parental education on math score. The straight lines that connect the vertical lines representing the size of the coefficients take on an upward trend, but the overall line is not exactly a straight line.

Table 3.18 shows results from the regression that uses the parental education variable as an interval variable. Math scores increase 2.78 units for each unit increase in parental education. The t value for the coefficient is 55.60, which is far beyond the

Table 3.17	Linear Regression of Parental Education on Math Score		
Independent Variable	**B**	**SE**	**t**
HS Only	2.43	.32	7.59
2-Yr. Degree	3.49	.33	10.57
4-Yr. Degree	8.10	.32	25.31
MA/MS	10.38	.35	29.66
PhD/MD	13.60	.39	34.87
Intercept	45.99	.29	—

Figure 3.3   Dummy Variable Coefficients for Parental Education Effects on Math Score With Line for Linear Effect

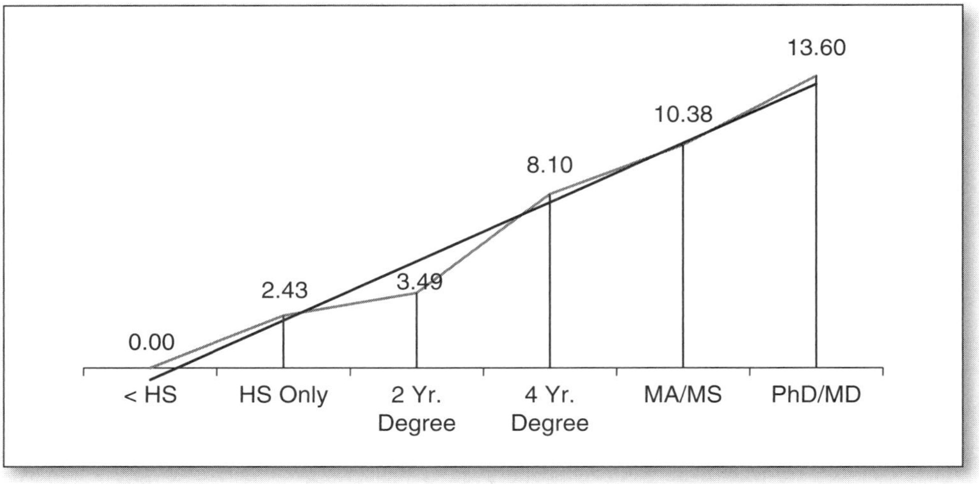

Table 3.18   Linear Regression of Parental Education on Math Score

Independent Variable	B	SE	t
Parental Ed.	2.78	.05	55.60
Intercept	42.49	.18	—

1.96 needed for statistical significance at the .05 level. A general interpretation of this coefficient is that higher parental education leads to higher math scores. A more specific interpretation is that each additional level of parental education leads to 2.78 more in mean math scores.

In Figure 3.3, a line is drawn that represents the linear coefficient in the earlier regression model. The line connecting the mean estimates from the dummy variable regression analysis generally follows the line representing the linear coefficient, but the dummy variable line does fluctuate around the linear coefficient line noticeably. Modeling parental education with a linear coefficient works well. If the dummy variable line was a U-shape or inverted U-shape, then using a linear coefficient to model the variable would not be appropriate.

Table 3.19 shows the differences between dummy variable coefficients for parental education. The unweighted mean for these differences is 2.72. Note that this mean does not take into account the size of the categories. Thus, this estimate will almost

Table 3.19 Differences in Dummy Variable Coefficients for Effects of Parental Education on Math Score Compared With the Linear Coefficient

$Coeff_2 - Coeff_1$	Difference	Linear B
HS Only $-$ < HS	2.43	2.78
2-Yr. Degree $-$ HS Only	1.06	2.78
4-Yr. Degree $-$ 2-Yr. Degree	4.61	2.78
MA/MS $-$ 4-Yr. Degree	2.28	2.78
PhD/MD $-$ MA/MS	3.22	2.78

always be different from the linear coefficient. The table also shows the linear B coefficient, which essentially forces the difference between category means to be equal. The example illustrates how the linear coefficient can be described as being similar to a mean of the differences in the means between categories.

The advantage of using the linear coefficient is that the linear coefficient is more parsimonious. The preference for parsimony in science is the idea that simpler explanations are preferred over more complex explanations if both explanations are equally effective. In this example, using the linear coefficient is more parsimonious since it captures the relationship with one coefficient rather than with seven coefficients. The effectiveness of the linear coefficient depends on how well it models the actual relationship between variables.

Linear Coefficient in Logistic Regression

We can also estimate linear coefficients in logistic regression. Table 3.20 shows the log odds of being in the top 25% on math for each parental education category. The results show that as parental education increases, the log odds of being in the top 25% in math increase. The last column in the table highlights this pattern by showing the difference between the log odds for the less-than-high-school category and all the other categories. The differences become less negative going from one category to the next. We can see that the relationship is basically linear since the coefficients increase as education increases. Increase in this case means that the coefficients become less negative.

In Table 3.21, I again examine linearity by first using dummy variables to consider the relationship between parental education and being in the top 25% in math. I do this by including dummy variables for parental education for all categories except the less-than-high-school category. Notice that the coefficients for the dummy variables in the logistic regression are the same as the differences between log odds shown in Table 3.20. The intercept is the log odds for the less-than-high-school category.

Table 3.20   Log Odds of Top 25% in Math Score by Parental Education

Parental Ed.	Log Odds Top 25% Math	Difference from < HS
< HS	−2.39	.00
HS Only	−1.86	.53
2-Yr. Degree	−1.65	.74
4-Yr. Degree	−.71	1.68
MA/MS	−.28	2.11
PhD/MD	.20	2.59

Table 3.21   Logistic Regression of Parental Education on Top 25% in Math

Independent Variable	B	SE	t
HS Only	.53	.12	4.42
2-Yr. Degree	.74	.12	6.17
4-Yr. Degree	1.68	.12	14.00
MA/MS	2.11	.12	17.58
PhD/MD	2.59	.13	19.92
Intercept	−2.39	.11	—

Next, I estimate the relationship between parental education and the top 25% in math by using an interval variable and a linear coefficient. The coefficient is significant at the .05 level and indicates that each additional unit of parental education results in a .55 higher log odds of top 25% in math.

Table 3.22   Logistic Regression of Parental Education on Top 25% in Math

Independent Variable	B	SE	t
Parental Ed.	.55	.01	55.00
Intercept	−3.04	.05	—

We can interpret the coefficient for parental education in the model that uses the interval variable as something like an average of the differences in log odds of top 25% in math between the categories for parental education. Table 3.23 shows the log odds for each category and the change in log odds as one goes from a lower parental education category to a higher category. The unweighted mean of the changes in log odds is 2.59 divided by 5, which equals .52. The coefficient for the interval parental education variable was .55. Although the mean of the changes is a rough approximation, it does provide a way to conceptualize the meaning of the linear coefficient. The regression takes the number of units with each value into account, whereas the estimate does not.

Table 3.23	Differences in Dummy Variable Coefficients for Effects of Parental Education on Top 25% in Math Compared With the Linear Coefficient	
Parental Ed.	**Log Odds Top 25% Math**	**Change in Log Odds**
< HS	−2.39	—
HS Only	−1.86	.53
2-Yr. Degree	−1.65	.21
4-Yr. Degree	−.71	.94
MA/MS	−.28	.43
PhD/MD	.20	.48

When an interval variable has a limited number of values, the variable can be modeled by using dummy variables. Using dummy variables for modeling an interval variable is a straightforward method for seeing the extent of a linear relationship.

Using Dummy Variables for a Continuous Variable

In the examples presented so far, when considering linear coefficients, the interval variables have had a limited number of values. Interval variables like this are called discrete interval variables. Another type of interval variable is called a continuous interval variable. This is a variable that could possibly take on all values on the number line within the range between the lowest and highest possible values.

In addition to the parental SES quartile variable in the HSLS that we used earlier, there is also a continuous parental SES variable. This variable is a standardized combination of parental education level, family income, and parental occupational status. The variable ranges from −1.75 to 2.28.

Table 3.24 shows the results of a logistic regression that uses parental SES as an independent variable and estimating a linear coefficient. The coefficient of 1.20 means

Table 3.24	Logistic Regression of Parental Education on Attends Private School		
Independent Variable	B	SE	t
Parental SES	1.20	.03	40.00
Intercept	−1.98	.02	—

that for every one unit of parental SES, the log odds of attending private school increases by 1.20 units. The linear coefficient is significantly different from zero indicating that there is a significant linear relationship.

Sometimes a linear coefficient is not significant. However, there still may be a relationship between the independent and the dependent variable. For example, the relationship between age in years and some dependent variables takes on a U-shape or an inverted U-shape.

One way to check in a straightforward manner is to divide the continuous interval variable into categories based on quartiles (4ths), quintiles (5ths), or deciles (10ths). In Table 3.25, I divided parental SES into deciles. The constraint in doing something like this is that there needs to be 25 or more cases in each category. With 20,133 cases in this analysis, I can use deciles since there are more than 2,000 cases in each decile.

Table 3.25	Logistic Regression of Parental Education on Attends Private School		
Independent Variable	B	SE	t
SES Decile 2	0.57	.17	3.35
SES Decile 3	1.01	.16	6.31
SES Decile 4	1.23	.16	7.69
SES Decile 5	1.50	.15	10.00
SES Decile 6	1.76	.15	11.73
SES Decile 7	2.15	.15	14.33
SES Decile 8	2.41	.15	16.07
SES Decile 9	2.75	.14	19.64
SES Decile 10	3.22	.14	23.00
Intercept	−3.55	.14	—

Figure 3.4   Dummy Variable Coefficients for SES Decile Effects on Attends Private School With Line for Linear Effect

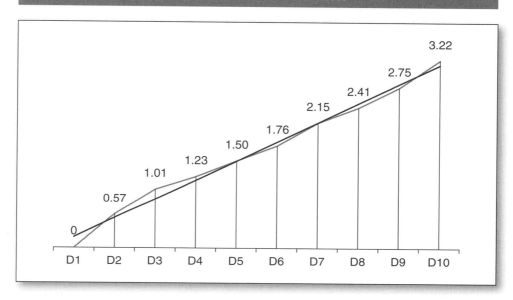

In Figure 3.4, I have graphed the dummy variable coefficients from the logistic regression for the effect of parental SES decile on the log odds of attending private school. The increase in the log odds of attending private school increases for the lowest deciles, flattens out a bit, and increases again for the highest deciles.

There are more formal ways of testing for nonlinearity such as adding a squared variable to the model. For parental SES, one would calculate a new variable that is parental SES squared. Parental SES and parental SES squared would be included in the regression model. If parental SES squared is significantly different from zero, then that is evidence of a nonlinear relationship. However, the advantage of using dummy variables is that the results are more straightforward to interpret. Later in the book, I discuss another way to examine linearity, which is to use spline variables.

Summary

Most introductory discussions of linear regression often start by explaining how to fit a line through a scatterplot of points. In this book, I start the discussion of linear regression with dummy variable regression. I show that a linear regression with a single dummy independent variable produces the same result as the *t* test for the difference in means between the two groups represented by the dummy variable. Dummy variables can also be used to capture a set of differences among three or more groups. A key analytical decision when there are three or more groups represented by a variable is which of the groups will be chosen as the key contrast group. Although linear regression involving dummy independent variables can be viewed as estimating differences in means between groups, logistic regression involving dummy independent variables can be viewed as estimating differences in log odds between groups.

When a variable is at the ordinal or interval level of measurement, the coefficients for dummy independent variables can be examined to determine the extent to which the effects of the variable are linear. A linear effect in the dummy variable context means as one goes from one category of the independent variable to the next, the decrease or increase in the dummy variable coefficient is constant. Thus, a coefficient for an interval variable in a linear regression or a logistic regression can be viewed as a constrained effect that forces the decreases or increases in the dependent variable as one goes from one value of the interval variable to the next to be equal.

Key Concepts

t test: the difference between sample means divided by an estimate of the standard deviation of the distribution of sample means.

dummy variable: a variable that has only two values with one value being coded as "1" and the other value being coded as "0."

linear regression: a statistical technique that measures the relationship between independent variables and an interval dependent variable; linear regression can be viewed as analyzing differences between means.

logistic regression: a statistical technique that measures the relationship between independent variables and a dependent variable measured by a dummy variable; logistic regression can be viewed as analyzing differences between log odds.

linear effect: an increase/decrease in the dependent variable that is constant as the independent variable increases unit by unit.

Chapter Exercises

1. Replicate the table of dummy variable coefficients in linear regression where the independent variable is family structure (STEP, SINGLE, FAMOTH) and the dependent variable is math score (X2TXMTSCOR).

2. Replicate the table of dummy variable coefficients in logistic regression where the independent variable is family structure and the dependent variable is top 25% in math score (HIGHMATH).

3. Create a table of dummy variable linear regression coefficients where the independent variable is family structure and the dependent variable is SES (X2SES). Which family structure categories are highest in SES, and which are lowest?

4. Create a table of dummy variable logistic regression coefficients where the independent variable is family structure and the dependent variable is private high school (PRIVATE). Which family structure categories are most likely to attend private high school? Are any groups equal?

5. Plot the relationship between SES quintile score (SESQ52, SESQ53, SESQ54, SESQ55) and private high school by using dummy variable logistic regression coefficients. Is the line a straight line?

4

Key Regression Modeling Concepts

Unit Vector: Estimating the Intercept

The intercept in dummy variable regression can come across as a magical thing. Without the researcher doing anything other than changing the variables in the model, the mean for the excluded category is estimated by the regression, seemingly without request. The reality behind this magic is that the researcher actually does enter a variable that estimates the mean for the excluded category.

The explanation for how this occurs involves thinking in more depth about the "a" in the regression equation:

$$Y = a + b_1X_1 + b_2X_2 + e \tag{1}$$

Two variables are provided in Equation 1, and a "b" coefficient is estimated for each one. However, some econometricians and statisticians would write the regression model as shown in Equation 2. This equation includes b_1 in place of a:

$$Y = b_1 + b_2X_1 + b_3X_2 + e \tag{2}$$

Other econometricians and statisticians would write the model in a third, fully compatible, manner. This equation includes b_1U rather than a or b_1 alone:

$$Y = b_1U + b_2X_1 + b_3X_2 + e \tag{3}$$

In Equation 3, the variable U is a column of ones. Sometimes U is referred to as the unit vector.

So if "a" is not a special coefficient but is a "b" coefficient applied to the unit vector, the next question is, Why enter the unit vector as a variable in the regression model? An answer to this question involves deeper consideration of the intercept. I will use an example where dummy variables for family structure are used to predict math scores with the following model:

$$Y = a + b_1X_{\text{STEP}} + b_2X_{\text{SING}} + b_3X_{\text{OTH}} + e \tag{4}$$

Following standard procedure, we leave out one category, two biological parents, and estimate the differences from that category by using dummy variables for the other three categories. The intercept a estimates the mean test score for those in two-biological-parent families:

$$Y = b_1U + b_2X_{\text{STEP}} + b_3X_{\text{SING}} + b_4X_{\text{OTH}} + e \tag{5}$$

We can rewrite Equation 4 by using b_1U instead of a. I use the unit vector to affirm that a is a coefficient like the b, and it is the coefficient for the unit vector.

I believe that understanding the intercept is a key to the proper interpretation of the other coefficients in a dummy variable regression model.[1] To facilitate this understanding, I write equations and include below the equations notation showing the data vector for the particular variable.

Nestedness

In the dataset that I am using for the examples, there are 20,133 cases. In Equation 6, this would result in a unit vector with 20,133 ones. There would also be that many zeros and ones for the three family structure variables. For purposes of illustration, I only show the four unique data records where each line is a data record:[2]

---

1. Hardy (1993) presents a thorough introduction to the use of dummy variables in regression. Hardy (1993), page 10, provides a useful discussion of factors to consider in choosing the contrast group.

2. Fox (2015) and Gordon (2010), page 214, use a table to present the coding for dummy variables in a manner similar to the matrix approach that this book uses.

$$Y = b_1 U + b_2 X_{\text{STEP}} + b_3 X_{\text{SING}} + b_4 X_{\text{OTH}} + e \tag{6}$$

$$\begin{bmatrix}1\\1\\1\\1\end{bmatrix}\quad\begin{bmatrix}0\\1\\0\\0\end{bmatrix}\quad\begin{bmatrix}0\\0\\1\\0\end{bmatrix}\quad\begin{bmatrix}0\\0\\0\\1\end{bmatrix}\qquad \begin{array}{l}\textbf{two bio.}\\ \text{bio./step.}\\ \text{single}\\ \text{other}\end{array}$$

Please note two things about the data. First, the record for those who are in two-biological-parent families is 0 for the three other family structure variables but 1 on the unit vector. Second, the vector for each family structure category is "nested" in the unit vector, meaning that the vector is a subset of the unit vector.

The next diagram with Equation 7 shows a model where the biological-parent/stepparent category is the excluded category, and as a result, the record for those in a biological-parent/stepparent family is 0 on the other three family structure variables but is 1 on the unit vector. The records for those in the three other categories are 1 for that particular family structure category and 1 on the unit vector:

$$Y = b_1 U + b_2 X_{\text{BIO}} + b_3 X_{\text{SING}} + b_4 X_{\text{OTH}} + e \tag{7}$$

$$\begin{bmatrix}1\\1\\1\\1\end{bmatrix}\quad\begin{bmatrix}1\\0\\0\\0\end{bmatrix}\quad\begin{bmatrix}0\\0\\1\\0\end{bmatrix}\quad\begin{bmatrix}0\\0\\0\\1\end{bmatrix}\qquad \begin{array}{l}\text{two bio.}\\ \textbf{bio./step.}\\ \text{single}\\ \text{other}\end{array}$$

The diagrams that follow Equations 8 and 9 show what the data would look like, first, when the variable for single-parent family is excluded and, second, when the variable for other family is excluded. In each case, the record for those in the excluded category is 0 for all the family structure variables and 1 on the unit vector. Remember that in dummy variable linear regression models like these, the intercept estimates the mean for the excluded category:

$$Y = b_1 U + b_2 X_{\text{BIO}} + b_3 X_{\text{STEP}} + b_4 X_{\text{OTH}} + e \tag{8}$$

$$\begin{bmatrix}1\\1\\1\\1\end{bmatrix}\quad\begin{bmatrix}1\\0\\0\\0\end{bmatrix}\quad\begin{bmatrix}0\\1\\0\\0\end{bmatrix}\quad\begin{bmatrix}0\\0\\0\\1\end{bmatrix}\qquad \begin{array}{l}\text{two bio.}\\ \text{bio./step.}\\ \textbf{single}\\ \text{other}\end{array}$$

$$Y = b_1 U + b_2 X_{\text{BIO}} + b_3 X_{\text{STEP}} + b_4 X_{\text{SING}} + e \tag{9}$$

$$\begin{bmatrix} 1 \\ 1 \\ 1 \\ \mathbf{1} \end{bmatrix} \quad \begin{bmatrix} 1 \\ 0 \\ 0 \\ \mathbf{0} \end{bmatrix} \quad \begin{bmatrix} 0 \\ 1 \\ 0 \\ \mathbf{0} \end{bmatrix} \quad \begin{bmatrix} 0 \\ 0 \\ 1 \\ \mathbf{0} \end{bmatrix} \qquad \begin{matrix} \texttt{two bio.} \\ \texttt{bio./step.} \\ \texttt{single} \\ \mathbf{\texttt{other}} \end{matrix}$$

So how does the regression program know which mean to assign to the coefficient for the unit vector when the unit vector is all ones in every case and does not seem to provide information that varies across models? The answer to this question involves considering a model that researchers never estimate but is a model that provides insight into interpreting the unit vector.

This model is the "intercept-only" model and is shown in Equation 10. The model includes only the unit vector and does not include any of the four dummy variables for family structure:

$$Y = b_1 U + e \tag{10}$$

$$\begin{bmatrix} 1 \\ 1 \\ 1 \\ 1 \end{bmatrix}$$

If one were to use a standard statistical package to estimate this model, one would have to select the option to suppress the intercept and then add the unit vector as an independent variable. This model with only the intercept will estimate the mean for all categories combined. Researchers do not estimate a model like this since generally social science research hypotheses involve differences in means between groups and not the size of the overall mean. However, we can view the intercept-only model in regression analysis as a starting point. We add additional variables to a regression model to see whether we can explain more variation than a model that estimates only the overall mean.

I suggest that the intercept-only model can be viewed as the starting point in a regression analysis. Adding dummy variables allows for estimation of differences from the mean estimated by the coefficient for the unit vector. However, the coefficient for the unit vector estimates the mean for only those groups not captured by the dummy variables in the model. I can illustrate this idea by showing the four possible dummy variable models that can be constructed with dummy variables for family structure, as follows:

Two-biological-parents category as contrast:

$$Y = b_1 U + b_2 X_{\text{STEP}} + b_3 X_{\text{SING}} + b_4 X_{\text{OTH}} + e \tag{11}$$

Biological-parent/stepparent category as contrast:

$$Y = b_1 X_{BIO} + b_2 U + b_3 X_{SING} + b_4 X_{OTH} + e \qquad (12)$$

Single-parent category as contrast:

$$Y = b_1 X_{BIO} + b_2 X_{STEP} + b_3 U + b_4 X_{OTH} + e \qquad (13)$$

Other-family category as contrast:

$$Y = b_1 X_{BIO} + b_2 X_{STEP} + b_3 X_{SING} + b_4 U + e \qquad (14)$$

The unit vector plays a different role in each model. Replacing the dummy variable for the contrast group with the unit vector forces the regression to estimate the mean for that group and to estimate differences from that mean for the other groups. The intercept estimates a mean. It estimates the mean for the groups not covered by the dummy variables nested in the unit vector. Although it is true that the intercept estimates the mean for the "excluded" category, we can also say that it estimates the mean for the "remainder" category.

Again, Table 4.1 shows the means for family structure.

Table 4.1	Mean Math Score by Family Structure
Family Structure	**Mean Math Score**
Two Bio.	52.88
Bio./Step.	50.32
Single	50.14
Other Fam.	49.10

The general rule when dummy variables are nested in the unit vector is that the coefficient for the unit vector estimates the mean for the excluded category. Notice in the regression in Table 4.2 how the mean estimated by the intercept changes as the excluded dummy variable changes.

Higher Order Differences

Although the concept of "nestedness" is useful for understanding the nature of the intercept, nesting variables has many other practical uses in regression analysis.

In the model in Equation 15, b_2 estimates the coefficient for those in biological-parent/stepparent families and b_3 estimates the coefficient for those in single-parent families. The comparison group for both groups is the two-biological-parent category. What alternatives are there for estimating the difference between the coefficients for

| Table 4.2 | Linear Regression of Family Structure on Math Score | | | |

Independent Variable	Model			
	1	2	3	4
	B	*B*	*B*	*B*
Two Bio.	—	2.56*	2.74*	3.78*
Bio./Step.	−2.56*	—	.18	1.22*
Single	−2.74*	−.18	—	1.04*
Other Fam.	−3.78*	−1.22*	−1.04*	—
Intercept	52.88	50.32	50.14	49.10
R^2	.021	.021	.021	.021

*$p < .05$.

those in biological-parent/stepparent families and those in single-parent families, the difference between b_2 and b_3?

$$Y = b_1 U + b_2 X_{STEP} + b_3 X_{SING} + b_4 X_{OTH} + e \tag{15}$$

$$\begin{bmatrix} 1 \\ 1 \\ 1 \\ 1 \end{bmatrix} \quad \begin{bmatrix} 0 \\ 1 \\ 0 \\ 0 \end{bmatrix} \quad \begin{bmatrix} 0 \\ 0 \\ 1 \\ 0 \end{bmatrix} \quad \begin{bmatrix} 0 \\ 0 \\ 0 \\ 1 \end{bmatrix} \quad \begin{matrix} \texttt{two bio.} \\ \textbf{\texttt{bio./step.}} \\ \textbf{\texttt{single}} \\ \texttt{other} \end{matrix}$$

One alternative is to reestimate the regression with the biological-parent/step-parent category as the contrast group. A major drawback to this approach is that it is not as parsimonious, or not as simple, as other alternatives. In particular, this approach involves estimating three extra coefficients that are not needed. In research, it is better to do as few statistical tests as possible to lower the risk of finding significant effects simply by chance.

One can get an estimate of the difference with the addition of only one estimated coefficient by creating a combined variable.[3] The combined variable is created by adding

---

3. Another way to test the difference between two coefficients is to estimate one model that includes the two variables in question and estimates the coefficients for those two variables and a second model that estimates one coefficient for a variable that is created by adding the two variables

together the biological-parent/stepparent variable and the single-parent variable. The model in Equation 16 replaces the biological-parent/stepparent variable with the combined variable. The single-parent variable is now nested in the combined variable. That is, the single-parent variable is a subset of the combined variable:

$$Y = b_1 U + b_1 X_{\text{STEP+SING}} + b_2 X_{\text{SING}} + b_3 X_{\text{OTH}} + e \tag{16}$$

$$
\begin{bmatrix} 1 \\ 1 \\ 1 \\ 1 \end{bmatrix}
\quad
\begin{bmatrix} 0 \\ 1 \\ 1 \\ 0 \end{bmatrix}
\quad
\begin{bmatrix} 0 \\ 0 \\ 1 \\ 0 \end{bmatrix}
\quad
\begin{bmatrix} 0 \\ 0 \\ 0 \\ 1 \end{bmatrix}
\qquad
\begin{array}{l} \texttt{two bio.} \\ \textbf{\texttt{step. + single}} \\ \textbf{\texttt{single}} \\ \texttt{other} \end{array}
$$

I call this type of difference a "second-order" difference. An example of a "first-order" difference is the difference estimated by a coefficient for a dummy variable. This coefficient estimates the difference between the mean for the group represented by the dummy variable and the mean for the contrast group as estimated by the intercept. A second-order difference is a difference between differences. In the dummy variable case, the second-order difference is the difference between the two dummy variable coefficients, which are both first-order differences.

In Model 1 in Table 4.3, the coefficient for the biological-parent/stepparent variable is −2.56 and the coefficient for the single-parent variable is −2.74. These coefficients are clearly not equal. The next question that then must be addressed by the researcher is whether the coefficients are significantly not equal. We can get differences between sample coefficients as a result of sampling variation even when the population coefficients are equal.

The second model in Table 4.3 replaces the biological-parent/stepparent variable with the combined variable (bio./step. + single) and keeps the single-parent variable. Notice that the coefficient for the combined variable is −2.56 and is the same as the coefficient for bio./step. in Model 2. However, the coefficient for single parent changes to −.18, which is the difference between −2.56 and −2.74.

This amount, −.18, is what we would need to add to −2.56 to get to −2.74. This coefficient has a t value (not shown) of −.72 and is not significant at the .05 level. So we replace one variable, bio./step., with another, bio./step. + single, and the coefficient stays the same. We keep one variable the same, single, and the coefficient changes? What is going on?

(Demaris, 2004, pp. 94–95; Gordon, 2010, pp. 220–223; Pindyck & Rubinfeld, 1998, pp. 132–133; Treiman, 2009, pp. 147–149). An F test can be used to see whether adding the variables significantly reduced model fit. If model fit is significantly reduced, then the two coefficients are not equal. I believe the nesting procedure that I show in this chapter is a simpler way to test the difference between coefficients than the F test and statistically produces the same result.

	Table 4.3 Linear Regression of Family Structure on Math Score	

	Model	
	1	2
Independent Variable	*B*	*B*
Two Bio.	—	—
Bio./Step.	−2.56*	—
Bio./Step. + Single	—	−2.56*
Single	−2.74*	−.18
Other Fam.	−3.78*	−3.78*
Intercept	52.88	52.88
R^2	.021	.021

*$p < .05$.

At first glance, one might think that the model would estimate the difference between the mean for the biological-parent/stepparent + single-parent category and the mean for two-biological-parent category. Although this is not what happened, this would be the case if the variable for single was not included in the model.

Again, the general rule when using nested dummy variables is the coefficient for a dummy variable that is not a subset of any other dummy variable will estimate the difference for the groups involved less any groups captured by other variables nested in that variable. Since the dummy variable for single-parent family is included in the model, the coefficient for the dummy variable for the combined variable for biological-parent/stepparent and single-parent estimates the difference for the biological-parent/stepparent category. There is an important lesson to be learned from this example. Although the meaning of the coefficient for a variable depends on the variable itself, the meaning also depends on the other variables in the model.

It may seem that I have given too much attention to the interpretation of the intercept in dummy variable regression especially since we pay little attention to the intercept in most research. However, the discussion of the intercept has allowed me to introduce two key concepts in regression modeling.

The first key concept is that of nestedness. In dummy variable regression, dummy variables are nested within the unit vector. Anytime a difference is calculated in a regression model, one variable is nested in another variable in the model. Nestedness is a vital concept for understanding small and big models in control modeling and for understanding interactions.

In addition, adding two or more dummy variables together creates a new variable that encompasses the summed variables. Replacing one of the dummy variables that was summed up with the "summer variable" causes the regression program to estimate a more complex difference.

Thus, the second key concept is the concept of higher order differences. When we went from the standard dummy variable model to the model with one dummy variable nested within another, we went from a simpler to a more complex model. Although the same means were involved in each model, we went from estimating only first-order differences to estimating first-order differences and a more complex second-order difference.

Constraints

In addition to nestedness and higher order differences, a third key concept in regression modeling is that of constraints. One common constraint is setting two coefficients to be equal.

Equation 17 has separate coefficients for the biological-parent/stepparent variable and the single-parent variable. Equation 18 sets the constraint that the coefficients for the biological-parent/stepparent variable and the single-parent variable are equal, in this case, both equal to b_2. A straightforward way to get the regression program to estimate b_2 in the second equation is to not include the separate X_{STEP} and X_{SING} variables but to include the combined $X_{STEP} + X_{SING}$ variable. This is shown in Equation 19.

Equation 20 shows the setup for calculation of a second-order difference that was discussed previously. The difference between the Equations 19 and 20 is that $b_3 X_{SING}$ does not appear in Equation 19. Essentially, by not including X_{SING}, then b_3 in Equation 19 is set to zero:

$$Y = a + b_2 X_{STEP} + b_3 X_{SING} + b_4 X_{OTH} + e \tag{17}$$

$$Y = a + b_2 X_{STEP} + b_2 X_{SING} + b_4 X_{OTH} + e \tag{18}$$

$$Y = a + b_2 (X_{STEP} + X_{SING}) + b_4 X_{OTH} + e \tag{19}$$

$$Y = a + b_2 (X_{STEP} + X_{SING}) + b_3 X_{SING} + b_4 X_{OTH} + e \tag{20}$$

Table 4.4 shows three models. The first model includes the bio./step. and single variables. In the second model, the single variable is nested in the bio./step. + single variable. The coefficient for single in Model 2 captures the difference between the coefficients for the bio./step. variable and the single variable in Model 1. The difference is −.18, and the *t* value for this coefficient (not shown) is −.72, which is not significant at the .05 level.

The third model includes the bio./step. + single variable and does not include bio./step. or single. Model 3 constrains the bio./step. and single coefficients in Model 1 to be equal or, alternatively, constrains the variable for single in Model 2 to be zero. Since the −.18 in Model 2 is not significant, including both the bio./step. and single variables is not necessary, and Model 3 is the best model.

Independent Variable	Model		
	1	2	3
	B	*B*	*B*
Two Bio.	—	—	—
Bio./Step.	−2.56*	—	—
Bio./Step. + Single	—	−2.56*	−2.66*
Single	−2.74*	−.18	—
Other Fam.	−3.78*	−3.78*	−3.78*
Intercept	52.88	52.88	52.88
R^2	.021	.021	.021

Table 4.4   Linear Regression of Family Structure on Math Score

*$p < .05$.

Summary

The unit vector lurks in the background in regression analysis and gets little attention in the literature on regression analysis. Although the coefficient for the unit vector *a* is not typically analyzed in a regression analysis, explaining the role of the unit vector particularly in dummy variable regression introduces the key concepts of nestedness and first-order differences.

In dummy variable regression, each dummy variable is nested in the unit vector, which means each dummy variable is a subset of the unit vector. In dummy variable regression, nestedness for a dummy variable leads to the coefficient for the dummy variable estimating a first-order difference. In this case, the first-order difference is the difference in means between the group represented by the dummy variable and the contrast group represented by the unit vector. We can also use regression modeling to estimate a higher order difference. For example, we can obtain a second-order difference by including a dummy variable in the model, which is a subset of another dummy variable.

The concept of a constraint in regression modeling involves setting a particular coefficient to zero. In control modeling, constraints are set by setting the coefficients to zero in smaller models for coefficients that are estimated in larger models. In interaction modeling, constraints are set by setting the coefficients to zero in additive models for coefficients that are estimated in interaction models.

Key Concepts

unit vector: a variable used in regression analysis that consists of a column of ones.

nestedness: the situation where one variable is a subset of a second variable or when the variables in one regression model are a subset of the variables in a second regression model.

first-order difference: the estimate of the difference between means for two groups.

higher order difference: a difference in a difference between means.

constraint: involves setting a coefficient for a particular variable to a certain value; a common constraint in regression modeling is to set the coefficient for a particular variable equal to zero.

Chapter Exercises

1. Replicate models 1, 2, and 3 for the "constraints" example in Table 4.4 by using dummy variables for family structure (STEP, SINGLE, FAMOTH) and by using math score as the dependent variable (X2TXMTSCOR).

2. Create two new variables that are bio./step. + other family (STEPFAMOTH) and single + other family (SINGLEFAMOTH).

    Estimate the following two linear regression models with math score as the dependent variable and include these independent variables:

    Model 2a: bio./step. + other family, single, other family

    Is the coefficient for other family in this model significant? Interpret this coefficient.

    Model 2b: bio./step., single + other family, other family

    Is the coefficient for other family in this model significant? Interpret this coefficient.

3. Conduct a "constraints" analysis by using dummy variables for family structure and private (PRIVATE) as the dependent variable in logistic regression. Use STEPSINGLE, STEPFAMOTH, and SINGLEFAMOTH in the models.

    Estimate and interpret the following models:

    Model 2a: bio./step., single, other family

    Model 2b: bio./step. + single, single, other family

    Model 2b: bio./step. + other family, single, other family

    Model 2d: bio./step., single + other family, other family

5

Control
Modeling

Elementary Control Modeling

One of the fundamental problems that researchers address when using regression analysis is determining the degree to which an effect on a dependent variable that is associated with a particular independent variable occurs as a result of the relationship between that independent variable and other independent variables. In regression analysis, we call these other independent variables "control variables."[1]

Table 5.1 is a simple example of using a control variable in ordinary least-squares regression analysis. The dependent variable is math score. The independent variables include a dummy variable for attends private school or public school and a set of dummy variables for parental education with high school or less as the excluded category. The first model shows that those students in private school score 5.06 higher than those

---

1. Allison (1999), pages 16–19, provides an enlightening, introductory discussion of the issue of control in regression analysis.

	Model	
	1	2
Independent Variable	*B*	*B*
Private	5.06*	2.66*
2-Yr. Degree	—	1.32*
4-Yr. Degree	—	5.57*
Grad. Degree	—	8.72*
Intercept	50.80	47.88
R^2	.035	.140

Table 5.1 Control Model With Math Score as Dependent Variable

*$p < .05$.

students in public school. The coefficient in the second model for private is 2.66, and this coefficient decreased 47% from 5.06 in the first model.

The coefficient for private in the first model is the difference in mean math scores between those who attend private school and those who attend public school. The difference in the means is 5.06, and that is the same value as the coefficient for the dummy variable for private in Table 5.1.

When the three dummy variables for parental education are added to the model, the coefficient for private school decreased to 2.66. An important question for understanding control modeling is why does the coefficient decrease? I use two ways to explain what happens to the coefficient for the independent variable of interest when a control variable is added to a regression.[2]

The first way to explain how control works is to use the analytical technique of elaboration.[3] In this instance, elaboration involves considering the difference between the mean math score for those in private and public schools within categories of parental education.

This method "controls" for parental education by considering the mean difference in math scores between respondents in private and public schools only for those respondents who have parents with the same level of education. This method produces

2. An alternative way of thinking about control variables is to use the concepts of confounding, mediating, and suppressing variables as discussed in Demaris (2004), pages 98–104; Gordon (2010), Chapter 10; and Agresti and Finlay (2009), pages 307–313.

3. Linneman (2014), Chapter 10, discusses how to use elaboration to understand how control works in regression. The textbook also explains how to use small and big models to do control modeling.

four mean differences, one for each level of parental education. Each difference is thus calculated with parental education controlled.

Elaboration for Controlling

Table 5.2 shows the difference between those respondents in private and public schools in mean math scores within categories of parental education. Although the overall difference is 5.06, the difference in each parental education category is less. A rough estimate of how much this method for "controlling" for parental education lowered the mean differences is the mean of these differences, which is 2.95. The mean of the differences is close to the coefficient for private, 2.66, in the regression that controlled for parental education.

Table 5.2	Means for Math Score for School Control Within Parental Education		
Parental Education	**Private/Public**	**Mean Math**	**Difference**
< HS or HS Only	Public	47.79	3.98
	Private	51.77	
2-Yr. Degree	Public	49.09	3.67
	Private	52.76	
4-Yr. Degree	Public	53.55	2.24
	Private	55.79	
Grad. Degree	Public	56.84	1.91
	Private	58.75	
Mean Difference			2.95
Total	Public	50.80	5.06
	Private	55.86	

Demographic Standardization for Controlling

A second way to explain how control works is by using the method of demographic standardization.[4] In this method, an overall mean is viewed as a weighted sum of subgroup means. In direct demographic standardization, an overall mean is adjusted by changing the weights applied to the subgroup means.

---

4. Treiman (2009), Chapter 2, presents an excellent discussion of using elaboration and standardization for statistical control.

A classic example of direct demographic standardization is a standardized crude death rate. The distribution of deaths in human populations shows higher rates near birth and then lower rates for childhood, adolescence, and young adulthood. Death rates start rising in middle adulthood.

The crude death rate is the total number of deaths divided by the midyear population. We can view this rate as a weighted sum of death rates for age groups weighted by the size of the age groups. Two populations can differ in overall crude death rates as a result of underlying differences in age-specific death rates and in underlying differences in age structure. Populations with age structures that feature higher proportions in younger age groups will have a lower crude death rate from that influence. Populations that feature higher proportions in older age groups will have a higher crude death rate from that influence. Directly standardized crude death rates use a standard age distribution and by doing so "control" for the effect of age structure on the overall crude death rate.

The first step in considering the issue of control is to specify an independent variable of interest that is related to the dependent variable. In Table 5.3, we use school control as the independent variable of interest and we see that those in private schools score higher in math than those in public schools. The second step is to specify a control variable that, first, has an effect on the dependent variable and, second, is related to the independent variable of interest.

We will use parental education as a control variable in our consideration of the effect of school control on math scores. The rationale is that we know that parents with more education are more likely to send their children to private schools than parents with less education. We see in Table 5.3 that that those respondents with parents with more education score higher in math than those with parents with less education.

Table 5.3	Means for School Control and Parental Education

School Control	Math Mean
Public	50.80
Private	55.86
Parental Ed.	
< HS or HS Only	48.07
2-Yr. Degree	49.48
4-Yr. Degree	54.09
Grad. Degree	57.44

The second characteristic of an effective control variable is that the variable be related to the independent variable of interest. In Table 5.4, we see that those respondents in private school are more likely to have a parent who is a 4-year college graduate or has a graduate degree than those respondents in public school.

Table 5.4 Percentages for Type of School		
	Percentages	
Parental Ed.	**Public**	**Private**
< HS or HS Only	39.6	15.0
2-Yr. Degree	22.2	13.1
4-Yr. Degree	22.6	35.9
Grad. Degree	15.6	36.0
Total	100.0	100.0

Demographic standardization rests on the idea that the mean for any group is the weighted mean of the means for subgroups within the group. In this example, the means for respondents in public and private schools are the weighted means of the means in parental education subgroups.

Table 5.5 shows the public and private means expressed as weighted sums of the means for parental education subgroups. The product of the proportion in a subgroup times the mean for the subgroup is the contribution of the subgroup mean to the overall mean. Subgroups with higher proportions contribute more than subgroups with lower proportions.

The difference between the mean for respondents in private school and the mean for those in public school is 5.06. The difference is a result of two factors. One factor is that the respondents in private school have a higher math score in each category of parental education. The other factor is that the respondents in private school have higher proportions in the parental education categories where students score higher in math. The result from these two factors is a higher mean math score for those in private school compared with those in public school.

Table 5.5 shows the calculation of the mean for those respondents in private school and for those in public school as the weighted sum of the means in parental education subgroups. The calculation shows that the college graduate and graduate school categories contribute more to the mean for those in private school than for those in public school. On the other hand, the high school or less and some college categories contribute less.

Table 5.5   Group Means Expressed as Sum of Weighted Means for Subgroups

Parental Ed.	Public			Private		
	Prop.	Math Mean	Product	Prop.	Math Mean	Product
< HS or HS Only	.396	47.79	18.92	.150	51.77	7.77
2-Yr. Degree	.222	49.09	10.90	.131	52.76	6.91
4-Yr. Degree	.226	53.55	12.11	.359	55.79	20.03
Grad. Degree	.156	56.84	8.87	.360	58.75	21.15
Product Sum			50.80			55.86

In applying direct demographic standardization, the mean for those in private school is recalculated by using the parental education proportions for those in public school. Table 5.6 shows the calculation. The mean for those in private decreases from 55.86 to 53.99 when the parental education proportions for those in public school are used in place of the parental education proportions for those in private school.

Table 5.6   Calculation of Standardized Mean

Parental Ed.	Private			Private Standardized on Public		
	Prop.	Math Mean	Product	Prop.	Math Mean	Product
< HS or HS Only	.150	51.77	7.77	.396	51.77	20.50
2-Yr. Degree	.131	52.76	6.91	.222	52.76	11.71
4-Yr. Degree	.359	55.79	20.03	.226	55.79	12.61
Grad. Degree	.360	58.75	21.15	.156	58.75	9.17
Product Sum			55.86			53.99

The difference between those in private school and those in public school in mean math scores decreases from 5.06 (55.86 − 50.80) to 3.19 (53.99 − 50.80). Notice that the difference of 3.19 is fairly close in size to the coefficient for private in the regression model in Table 5.1, where parental education was controlled (2.66). Thus, demographic standardization produces an approximation to a control model in regression. The idea of controlling, then, is to constrain statistically two groups to have the same distribution on the control variable.

Small and Big Models

Professor Arthur Goldberger, a noted econometrician and teacher at the University of Wisconsin–Madison, referred to "short" and "long" models when discussing control models, and I use similar terms here.[5] A "big" model essentially adds more variables to the "small" model.

The simplest control model is shown in Table 5.7. A dummy variable for private school is the independent variable of interest. The control variables in the model include an interval variable measuring family income, a set of dummy variables measuring parental education (high school or less excluded), and a set of dummy variables measuring family structure of the respondent (two biological parents excluded). Those in private school score 1.78 higher in math scores than those in public school when family income, parental education, and family structure are controlled.

One way to think about this is to view the result as what the difference would be if those in private school and those in public school had the same distributions on family income, parental education, and family structure. We would use this one-model approach if we wanted to know the effect of the independent variable of interest when correlated, but theoretically less interesting factors are controlled. In this case, attendance at private school is related to family income, parental education, and family structure, but perhaps we are interested in the influence of attending private school itself and not interested in the characteristics of the children who attend private school. However, often we want to determine how much of the initial difference can be attributed to the control variables, and in this instance in Table 5.8, we use two regression models.

Table 5.8 shows that the effect for private school is 5.06 in the small model and 1.78 in the big model. The coefficient decreased 3.28 in actual magnitude for a decrease of 65%. Thus, more than one half of the difference between those in private school and those in public school in math scores is a result of differences in the distributions between the two groups on family income, parental education, and family structure:

---

5. Goldberger (1998) calls the models "short" and "long" because he has equations for regression models in mind. I call the models "small" and "big" having presentation tables in mind. A small model has a smaller number of coefficients in a column of a table, and a big model has a bigger number of coefficients.

Table 5.7	Linear Regression of Independent Variables on Math Score

Independent Variable	B
Private	1.78*
Family Income	.13*
2-Yr. Degree	1.14*
4-Yr. Degree	4.79*
Grad. Degree	7.34*
Bio./Step.	−1.33*
Single	−.82*
Other Fam.	−1.87*
Intercept	47.92
R^2	.157

*$p < .05$.

Table 5.8	Small and Big Models With Math Score as Dependent Variable

Independent Variable	Model	
	1	2
	B	B
Private	5.06*	1.78*
Family Income	—	.13*
2-Yr. Degree	—	1.14*
4-Yr. Degree	—	4.79*
Grad. Degree	—	7.34*
Bio./Step.	—	−1.33*
Single	—	−.82*
Other Fam.	—	−1.87*
Intercept	50.80	47.92
R^2	.035	.157

*$p < .05$.

$$\text{Change} = \frac{\text{big model coefficient} - \text{small model coefficient}}{\text{small model coefficient}} \times 100$$

$$= [(1.78 - 5.06)/5.06] \times 100 = 64.8\%$$

Family income, parental education, and family structure were effective control variables because the variables were strongly related to the dependent variable and strongly related to the independent variable of interest. In the regression in Table 5.8, those with higher family incomes had higher math scores than those not with high family incomes, those with higher parental education had higher math scores than those not with higher parental education, and those in two-biological-parent families had higher math scores than those not in two-biological-parent families.

Table 5.9 illustrates how family income, parental education, and family structure are strongly related to attending private school. Those in private school have higher

Table 5.9   Percentages for Control Variables by Type of School

Family Income ($)	Public	Private
0–35,000	30.2	10.5
36,000–75,000	33.5	21.5
76,000–115,000	19.0	23.7
116,000+	17.3	44.3
Total	100.0	100.0
Parental Ed.	**Public**	**Private**
< HS or HS Only	39.5	15.0
2-Yr. Degree	22.2	13.1
4-Yr. Degree	22.6	35.9
Grad. Degree	15.7	36.0
Total	100.0	100.0
Family Structure	**Public**	**Private**
Two Bio. Parents	53.7	72.3
Bio./Step.	15.1	8.3
Single Parent	21.7	14.3
Other Fam.	9.5	5.1
Total	100.0	100.0

family income and are more likely to have higher parental education and to be in two-biological-parent families than those in public school.

The big model in which family income, parental education, and family structure were added explained 65% of the effect for private that we found in the small model. The big model allows us to determine how much the coefficient for private changes when the three control variables were added, but it does not allow us to determine the role of each control variable in explaining the coefficient. To determine the role of each control variable, we will need to estimate a series of control models. Complicating this decision is the fact that the control variables are usually correlated with one another.

Allocating Influence With Multiple Control Variables

The Venn diagram that follows illustrates the problem of determining the relative influences of family income and parental education in explaining the effect of private school on math scores.[6] The circle for family income captures the part of the private school effect explained by family income, whereas the parental education circle does the same for parental education. The overlap captures the correlation between family income and parental education.[7]

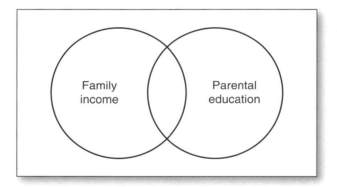

The problem in regard to control modeling is that whichever variable is added to the regression model first will capture the part of the private coefficient explained by the joint correlation of family income and parental education.[8] If we choose a modeling approach where we add variables in steps, the variable added earlier will have an

---

6. Agresti and Finlay (2009), pages 304–307, provide a brief discussion of control. They also use the idea of a Venn diagram to illustrate control, page 445.

7. The classic way to illustrate control is to use three-dimensional graphs such as in Fox (2015).

8. Goldberger (1998) points out that the first variable entered when doing a small–big model analysis always captures more of the R^2 and, therefore, allocating R^2 serves "no useful purpose."

advantage over the variable added later. If we add each control variable separately, then the joint influence is captured repeatedly in the models. Although there is no solution to this problem, we can understand the extent of the problem by modeling the control variables in different ways.

One-at-a-Time Without Controls

Table 5.10 shows the one-at-a-time model without controls approach. That is, each control variable is added without the other control variables in the model. The coefficient for attending private school decreased to 3.14 (−38%) when we added family income, to 2.66 (−47%) when we added parental education, and to 4.61 (−9%) when we added family structure. Model 5 includes all control variables, and the coefficient for private decreased to 1.78 (−65%).

Notice that the percentage change when all the control variables were added to the model is less than the sum of the percentage changes when each variable was added separately. The overlap shown in the Venn diagram is captured more than once.

Table 5.10 One-at-a-Time Without Controls With Math Score as Dependent Variable

Independent Variable	Model				
	1	2	3	4	5
	B	*B*	*B*	*B*	*B*
Private	5.06*	3.14*	2.66*	4.61*	1.78*
Family Income	—	.23*	—	—	.13*
2-Yr. Degree	—	—	1.32*	—	1.14*
4-Yr. Degree	—	—	5.57*	—	4.79*
Grad. Degree	—	—	8.72*	—	7.34*
Bio./Step.	—	—	—	−2.04*	−1.33*
Single	—	—	—	−2.30*	−.82*
Other Fam.	—	—	—	−3.25*	−1.87*
Intercept	50.80	49.01	47.88	51.91	47.92
R^2	.035	.084	.140	.049	.157

*$p < .05$.

Step Approach

The step approach in Table 5.11 involves adding one control variable, then an additional control variable, and then one more control variable.[9] Rather than comparing the models with the control variable with Model 1, in this approach, Model 2 is compared with Model 1, Model 3 with Model 2, and Model 4 with Model 3.

The coefficient for attending private school decreased to 3.14 (−38%) when we added family income, to 1.91 (−39%) when we added parental education, and to 1.78 (an additional −7%) when we added family structure. The explanatory power of parental education decreases when using this approach compared with the one-at-a-time without controls approach because the first variable added gets the overlap in explanatory power as illustrated in the Venn diagram. Thus, only family income gets the overlap in explanatory power.

In Table 5.12, we add the dummy variables for parental education first and then the interval variable for family income. The coefficient for attending private school

Table 5.11	**Step Model With Math Score as Dependent Variable**			

	Model			
	1	2	3	4
Independent Variable	*B*	*B*	*B*	*B*
Private	5.06*	3.14*	1.91*	1.78*
Family Income	—	.23*	.13*	.13*
2-Yr. Degree	—	—	1.16*	1.14*
4-Yr. Degree	—	—	4.93*	4.79*
Grad. Degree	—	—	7.51*	7.34*
Bio./Step.	—	—	—	−1.33*
Single	—	—	—	−.82*
Other Fam.	—	—	—	−1.87*
Intercept	50.80	49.01	47.26	47.92
R^2	.035	.084	.153	.157

*$p < .05$.

---

9. Demaris (2004), page 88, provides an example of a step model.

decreased to 2.66 (−47%) when we added parental education and to 1.91 (−28%) when we added family income. Family income explained 38% of the coefficient when it was added first compared with 28% when added second. This shows that order of entry definitely matters.

In the case of parental education and family income in Table 5.12, the variable entered first not only captures the joint influence with the other variable, but also it captures the joint influence with other possible explanatory variables, whether analyzed in the analysis or not.

Please note that the step model described earlier is not created by the "stepwise" procedure found in the SPSS statistical program and in other statistical programs. Here, the step model refers to starting with a small model and adding variables by steps to create larger models that answer theoretical questions. The stepwise procedure in SPSS includes in the first model the independent variable that adds the most to R^2, then includes in the second model the variable that adds the second most to R^2, and so on. Using the step approach described in this book requires the researcher to enter variables in an order that is dictated by theoretical considerations, not by statistical considerations.

Table 5.12 Step Model With Math Score as Dependent Variable

Independent Variable	Model			
	1	2	3	4
	B	B	B	B
Private	5.06*	2.66*	1.91*	1.78*
2-Yr. Degree	—	1.32*	1.16*	1.14*
4-Yr. Degree	—	5.57*	4.93*	4.79*
Grad. Degree	—	8.72*	7.51*	7.34*
Family Income	—	—	.13*	.13*
Bio./Step.	—	—	—	−1.33*
Single	—	—	—	−.82*
Other Fam.	—	—	—	−1.87*
Intercept	50.80	47.88	47.26	47.92
R^2	.035	.140	.153	.157

*$p < .05$.

One-at-a-Time With Controls

The objective of control modeling is to determine how much of the effect of the independent variable of interest can be explained by a control variable. We have observed how correlation between control variables complicates the analysis. A one-at-a-time model with controls deals with the correlation issue by considering the influence of each control variable with all other control variables in the model.

Model 1 determines the baseline effect for attending private school. To determine the influence of the control variables, Model 5 is compared with Models 2, 3, and 4. Models 2, 3, and 4 each contain two control variables, and Model 5 adds the third control variable. This allows us to consider the influence of each control variable above that of the other control variables.

In Table 5.13, the coefficient for attending private school decreased to 1.78 from 2.47 when family income was added (−28%), comparing Models 2 and 5. The coefficient for private school decreased to 1.78 from 2.90 when parental education was added (−39%), comparing Models 3 and 5. Finally, the coefficient decreased to 1.78 from 1.91 when family structure was added (−7%), comparing Models 4 and 5.

Table 5.13 One-at-a-Time With Controls With Math Score as Dependent Variable

Independent Variable	Model 1 B	2 B	3 B	4 B	5 B
Private	5.06*	2.47*	2.90*	1.91*	1.78*
Family Income	—	—	.22*	.13*	.13*
2-Yr. Degree	—	1.28*	—	1.16*	1.14*
4-Yr. Degree	—	5.37*	—	4.93*	4.79*
Grad. Degree	—	8.46*	—	7.51*	7.34*
Bio./Step.	—	−1.42*	−1.73*	—	−1.33*
Single	—	−1.08*	−1.57*	—	−.82*
Other Fam.	—	−2.14*	−2.55*	—	−1.87*
Intercept	50.80	48.62	49.95	47.26	47.92
R^2	.035	.145	.092	.153	.157

*$p < .05$.

Unlike the other approaches, the one-at-a-time with controls approach has a series of intermediate models that have less analytical value than the intermediate models in the other approaches. The one-at-a-time with controls allows us to affirm the results from the one-at-a-time without controls approach that family income and parental education both have strong explanatory power for explaining the effect for private school. The model also affirms that parental education has more explanatory power than family income.

The difference in the explanation between the one-at-a-time without controls approach and the one-at-a-time with controls approach lies in the handling of the part of the effect of the independent variable of interest explained by the joint correlation between variables. In the first approach, the explanation provided by the joint correlation is allocated to both variables, whereas in the second approach, the explanation is allocated to neither variable.

Hybrid Approach

All control variables do not play the same role in the analysis. Some control variables are simply correlated with the independent variable of interest. We need to control for

Table 5.14 Hybrid Approach With Math Score as Dependent Variable

Independent Variable	Model		
	1	2	3
	B	*B*	*B*
Private	5.06*	1.91*	1.78*
Family Income	—	.13*	.13*
2-Yr. Degree	—	1.16*	1.14*
4-Yr. Degree	—	4.93*	4.79*
Grad. Degree	—	7.51*	7.34*
Bio./Step.	—	—	−1.33*
Single	—	—	−.82*
Other Fam.	—	—	−1.87*
Intercept	50.80	47.26	47.92
R^2	.035	.153	.157

*$p < .05$.

these variables, but they are not high in theoretical interest. Other control variables may have direct influences on the independent variable of interest, or they may represent the intermediate mechanism by which the independent variable of interest influences the dependent variable.

The hybrid approach is similar to the step approach in that the variables are added in steps. The first step involves adding control variables that are correlated with the independent variable of interest. Later steps add control variables that are the focus of the analysis. In this way, the hybrid model resembles a one-at-a-time with controls model.

In Table 5.14, the independent variable of interest is attending private school. It is well known that those students who attend private school are more likely to come from families with higher incomes and have parents with higher levels of education than those who attend public school. Higher family income and higher parental education are both known to increase chances of attending private school. Model 1 in Table 5.14 includes only the variable for private. Model 2 includes family income and parental education. The coefficient for private decreases from Model 1 to Model 2 as the influences of family income and parental education are controlled but remains significantly different from zero.

Suppose the researcher wants to focus on family structure and believes that part of the reason attending private school influences math scores is that those who live in less advantageous family structures are less likely to attend private school. Family structure operates like family income and parental education in that these variables represent who goes to private school rather than what happens in private school.

Model 3 controls for family structure. The coefficient for private is 1.78 and still significantly different from zero. This represents a decrease from 1.91 in Model 2 and a modest 7% decrease in the private coefficient. Once family income and parental education are controlled, we find that family structure plays a relatively small role in explaining the overall effect of private school on math scores.

Nestedness and Constraints

We can use the concepts of nestedness and constraints to analyze what happens when we use control models in regression analysis. Nestedness is the idea that the variables in one model are a subset of the variables in another model. In the case of control models, variables in the small model are a subset of variables in the big model. The fact that the small model is nested in the big model allows us to use the F test for improvement of model fit to determine whether the additional variables significantly increased R^2.[10]

When we add one variable to the small model to obtain the big model, the t test for the added variable will produce the same result as the F test. However, if we wanted to test whether adding the three dummy variables for family structure, for example, significantly improved model fit, the F test would measure whether adding all three

---

10. Allen (1997), pages 113–117, discusses how to use the F test to test changes in model fit between nested models. Agresti and Finlay (2009), page 337, shows how to calculate an F test.

variables at once significantly improved model fit. Sometimes we might add three variables to a model, and only one has a significant *t* test. In this case, we do an *F* test to see whether the one significant variable in the set of three variables improved R^2 enough to warrant adding all three variables.

One approach to setting constraints is to set two coefficients equal. The small and big model approach to control modeling is a different way of setting constraints than setting coefficients equal. The small model is a constrained big model where certain coefficients are set to be equal to zero. We set this constraint by not adding the variables to the model.

Example Using Logistic Regression

Private schools in the United States are generally considered to be higher quality schools than public schools partly as a result of advantages in resources and teacher expertise. The analysis in Table 5.15 considers differences in private school attendance between non-Hispanic White students and Black students.

Table 5.15	Small and Big Models With Private High School as Dependent Variable	

	Model	
	1	2
Independent Variable	*B*	*B*
Black	−.37*	−.02
Other Race/Ethnicity	−.34*	−.14*
Family Income	—	.04*
2-Yr. Degree	—	.36*
4-Yr. Degree.	—	1.10*
Grad. Degree	—	1.24*
Bio./Step.	—	−.71*
Single	—	−.32*
Other Fam.	—	−.54*
Intercept	−1.47	−2.56
−2 log likelihood	17,999.3	15,859.1

*$p < .05$.

White and Black students differ on factors correlated with private school attendance. The regression analysis in Table 5.15 considers the degree to which Black/White differences in private school attendance are a result of differences in family income, parental education, and family structure. The analysis shows that Black students are significantly less likely to attend private school than White students with other factors not controlled. The second model in Table 5.15 shows that when family income, parental education, and family structure are controlled, the Black/White difference is no longer significant.

Table 5.16 Percentages for School Control by Independent Variables

	School Control		
	Public	**Private**	**Total**
Race/Ethnicity			
White	81.4	18.6	100
Black	86.4	13.6	100
Other	86.0	14.0	100
Family Income ($)			
0–35,000	93.6	6.4	100
36,000–75,000	88.7	11.3	100
76,000–115,000	80.2	19.8	100
116,000+	66.2	33.8	100
Parental Educ.			
< HS or HS Only	93.0	7.0	100
2-Yr. Degree	89.5	10.5	100
4-Yr. Degree	76.0	24.0	100
Grad. Degree	68.6	31.4	100
Family Structure			
Two Bio. Parents	78.9	21.1	100
Bio./Step.	90.1	9.9	100
Single	88.4	11.6	100
Other Fam.	90.3	9.9	100

A preliminary step in conducting a control modeling analysis is to examine a bivariate analysis for the relationship between the independent variables and the dependent variable and then to examine a bivariate analysis for the relationship between the independent variable of interest and the control variables. Table 5.16 shows percentages attending private school by race/ethnicity, family income, parental education, and family structure. The results show that Blacks have a smaller percentage attending private school than Whites. The control variables are also strongly related to the dependent variable. Those with higher family incomes are more likely to attend than those with lower family incomes, those with more educated parents are more likely to attend than those with less educated parents, and those in two-parent families are more likely to attend than those not in two-parent families.

Table 5.17	Percentages for Control Variables by Race/Ethnicity		
	Race/Ethnicity		
	White	Black	Other
Family Income ($)			
0–35,000	20.9	38.7	33.4
36,000–75,000	30.9	34.5	31.7
76,000–115,000	22.5	14.0	17.0
116,000+	25.7	12.8	17.9
Total	100	100	100
Parental Educ.			
< HS or HS Only	31.0	40.0	41.5
2-Yr. Degree	20.7	25.0	19.5
4-Yr. Degree	27.1	20.4	22.4
Grad. Degree	21.2	14.6	16.6
Total	100	100	100
Family Structure			
Two Bio. Parents	59.6	44.9	55.9
Bio./Step.	13.9	13.6	14.2
Single	18.8	28.9	20.6
Other Fam.	7.7	12.6	9.3
Total	100	100	100

Table 5.17 shows that the control variables are also related to the independent variable of interest. Blacks are less likely to have higher family incomes than are Whites, Blacks are less likely to have parents with a higher education than Whites, and Blacks are less likely to be in two-parent families than are Whites. The results of the bivariate analysis show that family income, parental education, and family structure are related to the dependent variable and that Blacks are disadvantaged on these variables compared with Whites. The results of the bivariate analysis suggest that controlling for family income, parental education, and family structure should explain part of the disadvantage that Blacks have compared with Whites in attending private school. The next question then is, to what degree do the control variables explain the disadvantage in attending private school for Blacks compared with Whites?

Table 5.18 shows the one-at-a-time model without controls. When we compare Models 2, 3, and 4 with Model 1, we see that although the coefficient for the

Table 5.18   One-at-a-Time Without Controls With Private High School as Dependent Variable

Independent Variable	Model				
	1	2	3	4	5
	B	B	B	B	B
Black	−.37*	−.13	−.19*	−.27*	−.02
Other Race/ Ethnicity	−.34*	−.21*	−.21*	−.32*	−.14*
Family Income	—	.06*	—	—	.04*
2-Yr. Degree	—	—	.42*	—	.36*
4-Yr. Degree.	—	—	1.41*	—	1.10*
Grad. Degree	—	—	1.78*	—	1.24*
Bio./Step.	—	—	—	−.89*	−.71*
Single	—	—	—	−.69*	−.32*
Other Fam.	—	—	—	−.89*	−.54*
Intercept	−1.47	−2.16	−2.48	−1.20	−2.56
−2 log-likelihood	17,999.3	16,639.8	16,660.5	17,606.9	15,859.1

$*p < .05$.

Black/White difference decreases when each control variable was added to the model, the decrease was largest for family income (−.37 to −.13). In fact, the Black/White difference was not significant when family income was controlled. However, the Black/White difference decreased by almost one half when parental education was controlled, so parental education is also an important explanatory factor (−.37 to −.19). In addition, the Black/White difference decreased by about one third when family structure was controlled so family structure also has noticeable explanatory power (−.37 to −.27). The drawback to this one-at-a-time without controls analysis is that family income, parental education, and family structure are related to one another so they share explanatory power and each one captures that shared explanatory power when added to the model separately.

Table 5.19 shows the one-at-a-time with controls approach. In this case, Models 2, 3, and 4 are compared with Model 5. Again, the biggest decrease in the Black/White coefficient occurs when family income is controlled. The Black

Table 5.19 One-at-a-Time With Controls With Private High School as Dependent Variable

Independent Variable	Model				
	1	2	3	4	5
	B	*B*	*B*	*B*	*B*
Black	−.37*	−.13	−.07	−.05	−.02
Other Race/Ethnicity	−.34*	−.20*	−.20*	−.15*	−.14*
Family Income	—	—	.06*	.04*	.04*
2-Yr. Degree	—	.41*	—	.36*	.36*
4-Yr. Degree.	—	1.34*	—	1.15*	1.10*
Grad. Degree	—	1.68*	—	1.30*	1.24*
Bio./Step.	—	−.76*	−.79*	—	−.71*
Single	—	−.43*	−.47*	—	−.32*
Other Fam.	—	−.65*	−.67*	—	−.54*
Intercept	−1.47	−2.23	−1.91	−2.77	−2.56
−2 log-likelihood	17,999.3	16,455.8	16,414.5	16,010.4	15,859.1

*$p < .05$.

coefficient in Model 2 is −.19 and decreases to −.02 in Model 5 for a decrease in the coefficient of .17. The Black coefficient also decreases when parental education is added, going from −.07 in Model 3 to −.02 in Model 5 for a decrease of .05. The decrease in the Black coefficient is smallest when family structure is added, going from −.05 in Model 4 to −.02 in Model 5 for a decrease of .03.

The analyses of the relative contributions of family income, parental education, and family structure that used the one-at-a-time without controls approach and the one-at-a-time with control approach were similar in saying the family income had the strongest explanatory power in explaining Black/White differences in attending private school. Both approaches showed that parental education was next in explanatory power followed by family structure. However, the relative amount of the Black coefficient explained by each variable was different in each approach as a result of the large amount of influence shared by the three factors.

Table 5.20 illustrates a step approach. The focus is on what role does family structure play in explaining the lower chances of attending private school for Black students compared with White students. First, family income and parental education are controlled since both variables are related to private school attendance and to family structure. Model 2 shows that the Black/White difference in private school attendance decreases to nonsignificance when family income and parental education

Table 5.20	Step Model With Private High School as Dependent Variable		
		Model	
	1	2	3
Independent Variable	*B*	*B*	*B*
Black	−.37*	−.05	−.02
Other Race/Ethnicity	−.34*	−.15*	−.14*
Family Income	—	.04*	.04*
2-Yr. Degree	—	.36*	.36*
4-Yr. Degree.	—	1.15*	1.10*
Grad. Degree	—	1.30*	1.24*
Bio./Step.	—	—	−.71*
Single	—	—	−.32*
Other Fam.	—	—	−.54*
Intercept	−1.47	−2.77	−2.56
−2 log-likelihood	17,999.3	16,010.4	15,859.1

*$p < .05$.

are controlled. Model 3 shows that adding family structure to the model that includes family income and parental education seems to lead to little further decrease in the Black/White coefficient.

The step model allows the variables added in the earlier steps to explain more of the coefficient for the independent variable of interest than the variables added in later steps as a result of the explanatory power of shared variation being captured in earlier steps. The contribution in the step model of adding family structure to the model in explaining the Black coefficient for Blacks of .03 [(−.02) − (−.05) = .03] is small compared with the joint contribution of family income and parental education [(−.05) − (−.37) = .32]. However, the contribution of family structure in the one-at-time without control, while the smallest of the three, was relatively larger. Thus, I suggest that a researcher explore different control modeling approaches so that the researcher has a firm idea about the relative contributions of the control variables used in the analysis under different circumstances before making conclusions.

Summary

Control modeling is the most widely used regression modeling approach that I discuss in this book. Control modeling starts with the concept of an independent variable of interest. The objective of regression modeling is to eliminate related influences that might explain the relationship, as measured by a regression coefficient, between the independent variable of interest and the dependent variable.

Standardization is a method used by demographers to hold constant the influence of a control variable on the relationship between an independent variable of interest and a dependent variable. Although demographic standardization does not produce the same result as using a control variable in dummy variable regression, the result from standardization is close enough that we can use the underlying concept in demographic standardization to understand control in regression analysis. That concept from demographic standardization is that control involves examining the difference between two groups on a dependent variable by making the underlying distribution on a third control variable equal for both groups.

A key underlying issue in control modeling is allocating the joint influence of two control variables in explaining the influence of an independent variable of interest on a dependent variable. The difficulty is there is no accepted method for allocating the joint influence of two control variables. If two variables are related to one another and to a dependent variable, then the first variable entered as a control variable will capture its unique influence and the joint influence that the variable shares with a second control variable.

The one-at-a time without controls approach to control modeling allows each control variable to capture its unique influence in explaining the effect of the independent variable of interest and any joint influences that it shares with any other control variable. On the other hand, the one-at-a time with controls approach allows each control variable only to capture its unique influence in explaining the effect of the independent variable of interest. The control variables added first in the step approach capture the joint influence of variables added in subsequent steps. Thus, the order that variables are added in the step approach has a great influence on the overall interpretation of the influence of control variables in explaining the independent variable of interest.

Researchers usually use only one control modeling approach to discuss their results in a research paper. However, I suggest that researchers carefully examine the results obtained by using alternative regression modeling approaches to understand the impact that adding control variables in some particular order had on the nature of the final result.

Key Concepts

control modeling: a modeling approach that involves first estimating the coefficient for an independent variable of interest and then adding control variables to take into account related influences.

elaboration: examining a relationship between two variables within categories of a third variable to control for the influence of the third variable.

demographic standardization: starts with the concept of a weighted mean where an overall mean is viewed as the sum of subgroup means weighted by the proportions for the subgroups; standardization involves creating an adjusted weighted mean for one group by using the subgroup proportions from a second group.

small and big models: a small model is a regression model where the variables included in the model are a subset of the variables included in a big model.

one-at-a-time without controls: a regression modeling procedure where only one control variable at a time is added to the small model, which includes the independent variable of interest.

step approach: a regression modeling procedure were one control variable is first added to the small model, which includes the independent variable of interest and then a second control variable is added to the second model to create a third model.

one-at-a-time with controls: a regression modeling procedure were only one control variable at a time is added to a smaller model, which includes the independent variable of interest and all other control variables.

hybrid approach: a regression modeling procedure that combines the step model regression modeling approach with the one-at-a-time with controls approach.

Chapter Exercises

1. Replicate the regressions and the table for the "one-at-a-time without controls" example in Table 5.10 by using X2TXMTSCOR, PRIVATE, FAMINC, TWOYR, FOURYR, GRAD, STEP, SINGLE, and FAMOTH.

2. Conduct a "one-at-a-time without controls" analysis like in Table 5.10, and create a table to present the results. Include a bivariate preliminary analysis in your answer. Use the dummy variable for two-parent family as the independent variable of interest and family income, 2-year degree, 4-year degree, graduate degree, and private as control variables. Use X2TXMTSCOR, TWOPAR, FAMINC, PAREDFOUR, PRIVATE, TWOYR, FOURYR, and GRAD in the analysis.

   What is the relationship between the independent variables and the dependent variable? What is the relationship between the independent variable of interest and the control variables? What percentage of the coefficient for the independent variable of interest was explained by controlling for all three sets of variables?

Use the following formula to calculate that percentage:

$$\frac{\text{big model coefficient} - \text{small model coefficient}}{\text{small model coefficient}} \times 100$$

What percentage of the coefficient for the independent variable of interest was explained by controlling for each set of variables separately?

3. Conduct a "hybrid" analysis like in Table 5.14, create a table to present the results, and describe your findings. Use top 25% in math as the dependent variable, and use the dummy variable for two-parent family as the independent variable of interest; add family income, 2-year degree, 4-year degree, and graduate degree in the second model, and add private in the third model as control variables. Use HIGHMATH, TWOPAR, FAMINC, PAREDFOUR, PRIVATE, TWOYR, FOURYR, and GRAD in the analysis.

What is the relationship between the independent variables and the dependent variable? What is the relationship between the independent variable of interest and the control variables? What percentage of the coefficient for the independent variable of interest was explained by controlling for each set of variables in this stepwise manner?

Modeling Interactions

Interactions as Conditional Differences

In additive regression models, the coefficients for dummy independent variables capture differences between groups on the dependent variable, whereas the coefficients for interval independent variables capture the additional contribution of each unit of the interval variable. Occasionally, researchers consider interactive effects between two independent variables on the dependent variable. This requires modeling interactions.

Table 6.1 shows the means for math scores. The pattern in math scores is that those with a college-graduate parent score higher than those not with a college-graduate parent and those living with two biological parents score higher than those not living with two biological parents. The general pattern for the differences on the parental education variable holds within family structure categories. At the same time, the general pattern for differences on the family structure variable also holds within categories of the parental education variable.

Table 6.2 shows differences in math scores between categories of the family structure variable within categories of the parental education variable. The two-biological-parent category is the contrast group. The means for biological-parent/stepparent category and the single-parent category are lower than the mean for the two-biological-parent

Table 6.1   Mean Math Scores by Family Structure and College-Graduate Parent

Family Structure	Not College Grad.	College Grad.	Total
Two Bio.	49.27	56.28	52.88
Bio./Step.	48.18	53.72	50.32
Single	48.19	54.44	50.14
Other Fam.	46.95	53.83	49.10
Total	48.59	55.54	51.63

Table 6.2   Differences in Mean Math Scores by Family Structure and College-Graduate Parent

Family Structure	Not College Grad.	College Grad.	Difference in Differences
Two Bio.	—	—	—
Bio./Step.	−1.09	−2.56	−1.47
Single	−1.08	−1.84	−.76
Other Fam.	−2.32	−2.45	−.13

category both among those not with a college-graduate parent and among those with a college-graduate parent.

The pattern of mean differences by family structure can be compared for the two groups by taking the difference in the differences. This calculation shows that the mean differences by family structure are more negative in the college-graduate-parent category than in the not-college-graduate-parent category. The differences in the differences show whether there is interaction. The idea of interaction, then, is simply that mean differences on the first independent variable are conditional on the second independent variable.

Another way to compare is to switch which variable is used as the "conditioning" variable. Since the parental education variable was the conditioning variable in Table 6.2, I use family structure as the conditioning variable in Table 6.3.

Table 6.3 shows that although those not with a college-graduate parent have a lower math score than those with a college-graduate parent, the difference is less in the biological-parent/stepparent and single-parent categories than in the two-biological-parent category. These differences in differences are shown in the last column.

Note that the differences in differences are the same for when I use the parental education variable as the conditioning variable as when I use the family structure

Family Structure	Not College Grad.	College Grad.	Difference in Differences
Two Bio.	—	7.01	—
Bio./Step.	—	5.54	−1.47
Single	—	6.25	−.76
Other Fam.	—	6.88	−.13

Table 6.3 Differences in Mean Math Scores by Family Structure and College-Graduate Parent

variable as the conditioning variable. Thus, the same set of differences in differences can be interpreted in two different ways depending on what variable is chosen as the conditioning variable.

Interactions Between Dummy Variables

The first step in considering interactions between dummy variables in regression is to examine the additive model shown in Equation 1. This model includes C_1, which is "1" if college graduate parent and "0" if not. The model also includes family structure variables for biological parent/stepparent, single parent, and other family. C_0, not-college-graduate parent, and F_1, two biological parents, are the excluded variables:

$$Y = a + b_1 C_1 + b_2 F_2 + b_3 F_3 + b_4 F_4 \tag{1}$$

The following matrices show the data matrices for the additive model. The essence of an additive model is that the effects of one variable are not conditioned on the values of a second variable. We could say the values on one variable are independent of the values of the second. The coefficient for the intercept is the mean for not-college-graduate/two biological parents:

$$
\begin{array}{ccccc}
U & C_1 & F_2 & F_3 & F_4 \\
\begin{bmatrix} 1 \\ 1 \\ 1 \\ 1 \\ 1 \\ 1 \\ 1 \\ 1 \end{bmatrix} &
\begin{bmatrix} 0 \\ 0 \\ 0 \\ 0 \\ 1 \\ 1 \\ 1 \\ 1 \end{bmatrix} &
\begin{bmatrix} 0 \\ 1 \\ 0 \\ 0 \\ 0 \\ 1 \\ 0 \\ 0 \end{bmatrix} &
\begin{bmatrix} 0 \\ 0 \\ 1 \\ 0 \\ 0 \\ 0 \\ 1 \\ 0 \end{bmatrix} &
\begin{bmatrix} 0 \\ 0 \\ 0 \\ 1 \\ 0 \\ 0 \\ 0 \\ 1 \end{bmatrix}
\end{array}
$$

Equation 2 includes interaction variables. We can create the interactions variable between two sets of dummy variables by multiplying all the variables in the first set by all the variables in the second set:

$$Y = a + b_1 C_1 + b_2 F_2 + b_3 F_3 + b_4 F_4 + b_5 C_1 F_2 + b_6 C_1 F_3 + b_7 C_1 F \qquad (2)$$

U	C_1	F_2	F_3	F_4	$C_1 F_2$	$C_1 F_3$	$C_1 F_4$
1	0	0	0	0	0	0	0
1	0	1	0	0	0	0	0
1	0	0	1	0	0	0	0
1	0	0	0	1	0	0	0
1	1	0	0	0	0	0	0
1	1	1	0	0	1	0	0
1	1	0	1	0	0	1	0
1	1	0	0	1	0	0	1

The interaction variables are nested in what we might call the additive variables. All three interaction variables are nested within C_1. In addition, $C_1 F_2$ is nested within F_2, $C_1 F_3$ is nested within F_3, and $C_1 F_4$ is nested within F_4. As you can see, the interaction variables are nested in two different additive variables. The effects of family structure depend on parental education, and the effects of parental education depend on family structure. Thus, there is a decision about what is the "conditioning" variable.

The differences as captured by the interaction variables can be viewed in two ways. They can be viewed as capturing differences in parental education effects by family structure or as capturing differences in family structure effects by parental education.

In addition to creating an equation that models differences in effects for groups, we can create an equation that models the effects within each group. Equation 3 estimates the effects of family structure within categories of parental education. The change from the previous equation is that rather than using the F_2, F_3, and F_4 variables, the equation includes $C_0 F_2$, $C_0 F_3$, and $C_0 F_4$:

$$Y = a + b_1 C_1 + b_2 C_0 F_2 + b_3 C_0 F_3 + b_4 C_0 F_4 + b_5 C_1 F_2 + b_6 C_1 F_3 + b_7 C_1 F_4 \qquad (3)$$

Equations 2 and 3 take different approaches to modeling interactions. Equation 2 estimates the effects of family structure for those not with a college-graduate parent with the F_2, F_3, and F_4 coefficients and then the difference from those effects for those with a college-graduate parent with the $C_1 F_2$, $C_1 F_3$, and $C_1 F_4$ coefficients. In contrast, Equation 3 estimates the effects of parental structure for those not with a college-graduate parent with the $C_0 F_2$, $C_0 F_3$, and $C_0 F_4$ coefficients and then the effects of family structure for those with a college-graduate parent with the C_1F_2, C_1F_3, and C_1F_4 coefficients. I refer to the first equation as a "standard interaction model" and to the second equation as a "within-group effects" model.[1]

We can see the relationship of the interaction variables in the within-group effects model in the following matrices. The C_0F_2, C_0F_3, and C_0F_4 variables and the C_1F_2, C_1F_3, and C_1F_4 are not nested within one another. What is happening is the F_2 variable has been split into two parts with C_0F_2 being those in both category C_0 and category F_2 and with C_1F_2 being those in both category C_1 and category F_2. Thus, in the within-group effects model, we split the F_2 variable into two parts, one for those in C_0 and one for those in C_1. The same is true for F_3 and F_4.

The within-group effects model is the model that contains the coefficients that the standard interaction model is testing the differences between.[2] The primary reason that researchers may have difficulty interpreting interaction coefficients in the standard interaction model is a lack of clarity about the underlying within-group effects model:

$$
\begin{array}{cccccccc}
U & C_1 & C_0F_2 & C_0F_3 & C_0F_4 & C_1F_2 & C_1F_3 & C_1F_4 \\[4pt]
\begin{bmatrix}1\\1\\1\\1\\1\\1\\1\\1\end{bmatrix} &
\begin{bmatrix}0\\0\\0\\0\\1\\1\\1\\1\end{bmatrix} &
\begin{bmatrix}0\\1\\0\\0\\0\\0\\0\\0\end{bmatrix} &
\begin{bmatrix}0\\0\\1\\0\\0\\0\\0\\0\end{bmatrix} &
\begin{bmatrix}0\\0\\0\\1\\0\\0\\0\\0\end{bmatrix} &
\begin{bmatrix}0\\0\\0\\0\\0\\1\\0\\0\end{bmatrix} &
\begin{bmatrix}0\\0\\0\\0\\0\\0\\1\\0\end{bmatrix} &
\begin{bmatrix}0\\0\\0\\0\\0\\0\\0\\1\end{bmatrix}
\end{array}
$$

Table 6.4 shows results from regression analysis that examines the effects of family structure and parental education on math scores. Model 1 is the additive model and shows that those with a college-graduate parent score higher in math than those not with a college-graduate parent. The model also shows that those living with two biological parents score higher in math than those not living with two biological parents. An additive model assumes that the effect of family structure is the same at all levels of parental education. For example, the disadvantage of living

---

1. Gordon (2010), pages 253–277, presents an alternative discussion of the standard interaction model using the concept of conditional means.

2. Demaris (2004), page 147; Jaccard (1990), pages 42–45; and Hardy (1993), pages 44–46, show how to calculate within-group effects by hand but do not show how to use dummy variables to estimate those coefficients. The disadvantage of simply adding coefficients to calculate within-group effects is that no standard error for the coefficient is created.

Table 6.4   Interactions Between Family Structure and College-Graduate Parent With Math Score as Dependent Variable

Independent Variable	Model		
	1	2	3
	B	B	B
College-Graduate Parent	6.65*	7.01*	7.01*
Bio./Step.	−1.70*	−1.09*	—
Single	−1.39*	−1.08*	—
Other Fam.	−2.43*	−2.32*	—
Not College Grad. × Bio./Step.	—	—	−1.09*
Not College Grad. × Single	—	—	−1.08*
Not College Grad. × Other Fam.	—	—	−2.32*
College Grad. × Bio./Step.	—	−1.47*	−2.56*
College Grad. × Single	—	−.76*	−1.84*
College Grad. × Other Fam.	—	−.13	−2.45*
Intercept	49.45	49.27	49.27
R^2	.124	.125	.125

*$p < .05$.

in a single-parent family would be the same despite whether the student's parent had a college degree.

Model 2 in Table 6.4 is the standard interaction model.[3] A common way of misinterpreting the standard interaction model is to discuss the model as one would for control models. An example of such a misinterpretation would be to say "the effect for college graduate parent increased going from Model 1 to Model 2 and the effects for family structure decreased when the interactions were controlled." Although the standard interaction model is created by adding variables to the additive model, what happens when those variables are added is definitely not like what happens when variables are added in control modeling.

To understand properly what is happening when the interaction variables are added to the additive model, we should first estimate Model 3, which is the within-group effects model. Model 3 in this example takes the family structure variables in

---

3. Linneman (2014), Chapter 12, provides a basic introduction to calculating and interpreting interactions in regression.

Model 1 and splits them into two sets of effects, one set for those not with a college-graduate parent and one for those with a college-graduate parent. Model 3 shows that the family structure effects are more negative for those with a college-graduate parent than for those not with a college-graduate parent. What Model 3 does not tell us is whether the two sets of coefficients are significantly different from one another. If the two sets are not significantly different, then Model 1 is the correct specification. If the two sets of coefficients are significantly different, then Model 3 is the correct specification.

Thus, Model 2 tests whether the family structure effects for those not with a college-graduate parent are different from the family structure effects for those with a college-graduate parent. Two coefficients for the interaction variables in Model 2 are significantly different from zero. This shows that the effects for family structure for those with a college-graduate parent are significantly more negative than the coefficients for those not with a college-graduate parent.

So far, we have examined the effects of family structure conditioned on parental education. The second way to consider interactions is to examine the effects of parental education conditioned on family structure as shown in Equation 4 and the following matrices:

$$Y = a + b_1 F_2 + b_2 F_3 + b_3 F_4 + b_4 C_1 + b_5 C_1 F_2 + b_6 C_1 F_3 + b_7 C_1 F_4 \qquad (4)$$

$$
\begin{array}{cccccccc}
U & F_2 & F_3 & F_4 & C_1 & C_1F_2 & C_1F_3 & C_1F_4 \\
\begin{bmatrix} 1 \\ 1 \\ 1 \\ 1 \\ 1 \\ 1 \\ 1 \\ 1 \end{bmatrix} &
\begin{bmatrix} 0 \\ 1 \\ 0 \\ 0 \\ 0 \\ 1 \\ 0 \\ 0 \end{bmatrix} &
\begin{bmatrix} 0 \\ 0 \\ 1 \\ 0 \\ 0 \\ 0 \\ 1 \\ 0 \end{bmatrix} &
\begin{bmatrix} 0 \\ 0 \\ 0 \\ 1 \\ 0 \\ 0 \\ 0 \\ 1 \end{bmatrix} &
\begin{bmatrix} 0 \\ 0 \\ 0 \\ 0 \\ 1 \\ 1 \\ 1 \\ 1 \end{bmatrix} &
\begin{bmatrix} 0 \\ 0 \\ 0 \\ 0 \\ 0 \\ 1 \\ 0 \\ 0 \end{bmatrix} &
\begin{bmatrix} 0 \\ 0 \\ 0 \\ 0 \\ 0 \\ 0 \\ 1 \\ 0 \end{bmatrix} &
\begin{bmatrix} 0 \\ 0 \\ 0 \\ 0 \\ 0 \\ 0 \\ 0 \\ 1 \end{bmatrix}
\end{array}
$$

The coefficient for C_1 in Equation 4 measures the effect of parental education for those living with two parents. The coefficients for $C_1 F_2$, $C_1 F_3$, and $C_1 F_4$ measure the additional effect of parental education in the other three categories.

Equation 5 shows the within-group effects model where the parental education effect is conditioned on family structure. It uses $C_1 F_1$ rather than C_1 as in Equation 4. The estimated coefficient for C_1 in Equation 4 and the coefficient for $C_1 F_1$ in Equation 5 are the same, and both capture the parental education effect for those with two biological parents. However, Equation 5 estimates the parental education effect for those in each other family structure type rather than the differences for those types:

$$Y = a + b_1F_2 + b_2F_3 + b_3F_4 + b_4C_1F_1 + b_5C_1F_2 + b_6C_1F_3 + b_7C_1F_4 \qquad (5)$$

$$
U \qquad F_2 \qquad F_3 \qquad F_4 \qquad C_1F_1 \qquad C_1F_2 \qquad C_1F_3 \qquad C_1F_4
$$

$$
\begin{bmatrix} 1 \\ 1 \\ 1 \\ 1 \\ 1 \\ 1 \\ 1 \\ 1 \end{bmatrix}
\begin{bmatrix} 0 \\ 1 \\ 0 \\ 0 \\ 0 \\ 1 \\ 0 \\ 0 \end{bmatrix}
\begin{bmatrix} 0 \\ 0 \\ 1 \\ 0 \\ 0 \\ 0 \\ 1 \\ 0 \end{bmatrix}
\begin{bmatrix} 0 \\ 0 \\ 0 \\ 1 \\ 0 \\ 0 \\ 0 \\ 1 \end{bmatrix}
\begin{bmatrix} 0 \\ 0 \\ 0 \\ 0 \\ 1 \\ 0 \\ 0 \\ 0 \end{bmatrix}
\begin{bmatrix} 0 \\ 0 \\ 0 \\ 0 \\ 0 \\ 1 \\ 0 \\ 0 \end{bmatrix}
\begin{bmatrix} 0 \\ 0 \\ 0 \\ 0 \\ 0 \\ 0 \\ 1 \\ 0 \end{bmatrix}
\begin{bmatrix} 0 \\ 0 \\ 0 \\ 0 \\ 0 \\ 0 \\ 0 \\ 1 \end{bmatrix}
$$

The C_1F_1, C_1F_2, C_1F_3, and C_1F_4 variables in the previous matrices are the variables that result when the C_1 variable is subdivided into parts for each family structure category. Table 6.5 shows the results for the within-group effects model that uses the subdivided variables. In Model 3, the parental education effects appear less positive for those not living with two biological parents. The standard interaction model, Model 2, shows that the parental education effects for those not living with two biological parents are

Table 6.5	Interactions Between Family Structure and College-Graduate Parent With Math Score as Dependent Variable		

	Model		
Independent	**1**	**2**	**3**
Variable	**B**	**B**	**B**
Bio./Step.	−1.70*	−1.09*	−1.09*
Single	−1.39*	−1.08*	−1.08*
Other Fam.	−2.43*	−2.32*	−2.32*
College-Graduate Parent	6.65*	7.01*	—
Two Bio. × College Grad.	—	—	7.01*
Bio./Step. × College Grad.	—	−1.47*	5.54*
Single × College Grad.	—	−.76*	6.25*
Other Fam. × College Grad.	—	−.13	6.88*
Intercept	49.45	49.27	49.27
R^2	.124	.125	.125

*$p < .05$.

significantly less than the parental education effect for those living with two biological parents. The significant interactions in Model 2 indicate that Model 3 is the correct specification for parental education effects, not Model 1.

I have shown that although there is one standard interaction model for the interaction between two sets of dummy variables, there are two different within-group effects models. In the examples in Table 6.5, one within-group effects model showed that family structure effects are greater for those living with a college-educated parent than for those not. The other within-group effects model showed that the parental education effect was more for those living with two biological parents than for those not.

These are two ways of addressing the same underlying issue. To say that the disadvantage of not living with two biological parents is more for those living with a college-educated parent is the same as saying that the advantage of living with a college-educated parent is less for those not living with two biological parents. Depending on which variable that we choose as the conditioning variable, we can interpret the interaction results in two ways. I reiterate that it is imperative to decide on which variable is the conditioning variable to interpret a standard interaction model properly.

In the previous discussion, I outlined two types of interaction models, the standard interaction model and the within-group effects model. Equation 6 is a third type of model that I refer to as the "all differences" interaction model. This model estimates only first-order differences.

The parental education variable has two categories, and the family structure variable has four categories. Thus, there are eight combinations of the two variables. The all-differences interaction model estimates the difference in means for seven of the categories from one of the categories, the one chosen as the excluded variable. In Equation 6, the C_0F_1 variable is excluded.

The data matrices for the equation include seven interaction variables and the unit vector. The interaction variables are nested only within the unit vector. The coefficients for the interaction variables estimate the difference in the mean for the group captured by the interaction variable and the mean for the contrast group. The all-differences interaction model is a type of additive model because the interaction variables are not nested:

$$Y = a + b_1 C_0F_2 + b_2 C_0F_3 + b_3 C_0F_4 + b_4 C_1F_1 + b_5 C_1F_2 + b_6 C_1F_3 + b_7 C_1F_4 \qquad (6)$$

U	C_0F_2	C_0F_3	C_0F_4	C_1F_1	C_1F_2	C_1F_3	C_1F_4
1	0	0	0	0	0	0	0
1	1	0	0	0	0	0	0
1	0	1	0	0	0	0	0
1	0	0	1	0	0	0	0
1	0	0	0	1	0	0	0
1	0	0	0	0	1	0	0
1	0	0	0	0	0	1	0
1	0	0	0	0	0	0	1

There is one more model that involves interactions between two set of dummy variables, and I refer to this model as the "all-means" model. In this example, the all-means model involves including all eight interaction variables capturing the interaction between parental education and family structure. The unit vector is not included. The coefficients in this model estimate the mean for each subgroup, hence, the name "all-means model." This is not a model that we would estimate for research purposes. However, the model is useful for purposes of understanding interactions and the role of the intercept.

When we look at the data matrices that follow Equation 7, we observe that none of the variables is nested within any other variable and that is why the model would estimate all means. Substituting the unit vector for any one of the interaction variables would lead to estimation of differences from the mean for the excluded category:

$$Y = b_1 C_0 F_1 + b_2 C_0 F_2 + b_3 C_0 F_3 + b_4 C_0 F_4 + b_5 C_1 F_1 + b_6 C_1 F_2 + b_7 C_1 F_3 + b_8 C_1 F_4 \qquad (7)$$

$$
\begin{array}{cccccccc}
C_0 F_1 & C_0 F_2 & C_0 F_3 & C_0 F_4 & C_1 F_1 & C_1 F_2 & C_1 F_3 & C_1 F_4
\end{array}
$$

$$
\begin{bmatrix} 1 \\ 0 \\ 0 \\ 0 \\ 0 \\ 0 \\ 0 \\ 0 \end{bmatrix}
\begin{bmatrix} 0 \\ 1 \\ 0 \\ 0 \\ 0 \\ 0 \\ 0 \\ 0 \end{bmatrix}
\begin{bmatrix} 0 \\ 0 \\ 1 \\ 0 \\ 0 \\ 0 \\ 0 \\ 0 \end{bmatrix}
\begin{bmatrix} 0 \\ 0 \\ 0 \\ 1 \\ 0 \\ 0 \\ 0 \\ 0 \end{bmatrix}
\begin{bmatrix} 0 \\ 0 \\ 0 \\ 0 \\ 1 \\ 0 \\ 0 \\ 0 \end{bmatrix}
\begin{bmatrix} 0 \\ 0 \\ 0 \\ 0 \\ 0 \\ 1 \\ 0 \\ 0 \end{bmatrix}
\begin{bmatrix} 0 \\ 0 \\ 0 \\ 0 \\ 0 \\ 0 \\ 1 \\ 0 \end{bmatrix}
\begin{bmatrix} 0 \\ 0 \\ 0 \\ 0 \\ 0 \\ 0 \\ 0 \\ 1 \end{bmatrix}
$$

Table 6.6 shows the results for the all-differences model and the all-means model. The all-differences model includes seven interaction variables and excludes the variable for not-college-graduate parent/two biological parents. The unit vector is included instead of this variable. As a result, the intercept is the mean for that category.

The all-means model includes the dummy variable for not-college-graduate parent/two biological parents in place of the unit vector and, thus, estimates all means. The dummy variable coefficients in Model 1 can be obtained by subtracting the means in Model 2 from mean for the not-college-graduate-parent/two-biological-parent group.

	Model	
	1	**2**
Independent Variable	*B*	*B*
Intercept	49.27*	—
Not College Grad. × Two Bio.	—	49.27
Not College Grad. × Bio./Step.	−1.09*	48.18
Not College Grad. × Single	−1.08*	48.19
Not College Grad. × Other Fam.	−2.32*	46.95
College Grad. × Two Bio.	7.00*	56.27
College Grad. × Bio./Step.	4.44*	53.71
College Grad. × Single	5.17*	54.44
College Grad. × Other Fam.	4.55*	53.82
R^2	.125	—

Table 6.6 Interactions Between Family Structure and College-Graduate Parent With Math Score as Dependent Variable

*$p < .05$.

Interactions Between Dummy Variables and an Interval Variable

Modeling interactions between a set of dummy independent variables and an interval-level independent variable is simpler than the case of interactions between two sets of dummy variables because there is only one within-group effects model to consider.[4]

---

4. This book does not discuss interactions between two interval variables. I find that these interactions are difficult to interpret. As an alternative, I suggest dividing into categories the interval variable that it is conditioned on and then using the procedure for interactions between dummy variables and an interval variable. For a brief discussion of interactions between interval variables, see Gordon (2010), and for more extensive treatments, see Jaccard (1990, 2001).

Equation 8 is the standard interaction model for the interaction between family structure and socioeconomic status (SES). F_1 is two biological parents, F_2 is biological parent/stepparent, F_3 is single parent, and F_4 is other family. SES is parental SES quartile and has four values:

$$Y = a + b_1F_2 + b_2F_3 + b_3F_4 + b_4\text{SES} + b_5F_2\text{SES} + b_6F_3\text{SES} + b_7F_4\text{SES} \tag{8}$$

U	F_2	F_3	F_4	SES	$F_2\text{SES}$	$F_3\text{SES}$	$F_4\text{SES}$
1	0	0	0	1	0	0	0
1	0	0	0	2	0	0	0
1	0	0	0	3	0	0	0
1	0	0	0	4	0	0	0
1	1	0	0	1	1	0	0
1	1	0	0	2	2	0	0
1	1	0	0	3	3	0	0
1	1	0	0	4	4	0	0
1	0	1	0	1	0	1	0
1	0	1	0	2	0	2	0
1	0	1	0	3	0	3	0
1	0	1	0	4	0	4	0
1	0	0	1	1	0	0	1
1	0	0	1	2	0	0	2
1	0	0	1	3	0	0	3
1	0	0	1	4	0	0	4

In Equation 8, the interactions are formed by multiplying each dummy variable times the interval variable. Examination of the data matrices that follow Equation 8 shows that the interaction variables are all nested within the SES variable. This means the coefficients for the interaction variables will estimate differences from the coefficient for SES. The coefficient for the SES variable will capture the SES effect for the subgroup not covered by the interactions, those in two-biological-parent families:

$$Y = a + b_1F_2 + b_2F_3 + b_3F_4 + b_4F_1\text{SES} + b_5F_2\text{SES} + b_6F_3\text{SES} + b_7F_4\text{SES} \tag{9}$$

Equation 9 for the within-group effects model uses $F_1\text{SES}$ rather than SES as in the standard interaction model. Examination of the following data matrices shows that the within-group effects model basically takes the interval SES variable and breaks it into four subparts, one for each family structure group:

$$
\begin{array}{cccccccc}
U & F_2 & F_3 & F_4 & F_1 SES & F_2 SES & F_3 SES & F_4 SES \\
\begin{bmatrix}1\\1\\1\\1\\1\\1\\1\\1\\1\\1\\1\\1\\1\\1\\1\\1\\1\end{bmatrix} &
\begin{bmatrix}0\\0\\0\\0\\1\\1\\1\\1\\0\\0\\0\\0\\0\\0\\0\\0\end{bmatrix} &
\begin{bmatrix}0\\0\\0\\0\\0\\0\\0\\0\\1\\1\\1\\1\\0\\0\\0\\0\end{bmatrix} &
\begin{bmatrix}0\\0\\0\\0\\0\\0\\0\\0\\0\\0\\0\\0\\1\\1\\1\\1\end{bmatrix} &
\begin{bmatrix}1\\2\\3\\4\\0\\0\\0\\0\\0\\0\\0\\0\\0\\0\\0\\0\end{bmatrix} &
\begin{bmatrix}0\\0\\0\\0\\1\\2\\3\\4\\0\\0\\0\\0\\0\\0\\0\\0\end{bmatrix} &
\begin{bmatrix}0\\0\\0\\0\\0\\0\\0\\0\\1\\2\\3\\4\\0\\0\\0\\0\end{bmatrix} &
\begin{bmatrix}0\\0\\0\\0\\0\\0\\0\\0\\0\\0\\0\\0\\1\\2\\3\\4\end{bmatrix}
\end{array}
$$

Model 3 in Table 6.7 is the within-group effects interaction model, and the SES coefficient is largest for two biological parents and smaller for the other three groups.

Table 6.7 Interactions Between Family Structure and Parental SES With Math Score as Dependent Variable

	Model		
	1	2	3
Independent Variable	B	B	B
Bio./Step.	−1.58*	.04	.04
Single	−.53*	.64	.64
Other Fam.	−1.59*	−1.24*	−1.24*
SES QUAR	3.32*	3.52*	—
Two Bio. × SES QUAR	—	—	3.52*
Bio./Step. × SES QUAR	—	−.64*	2.88*
Single × SES QUAR	—	−.51*	3.01*
Other Fam. × SES QUAR	—	−.11	3.41*
Intercept	43.75	43.23	43.23
R^2	.147	.147	.147

*p < .05.

Model 2 is the standard interaction model and shows that the SES coefficient for other family is not significantly different than the coefficient for two biological parents, whereas the coefficients for the other two groups are significantly smaller than the coefficient for two biological parents.

It is possible to estimate a third-order difference in a two-way interaction model. A first-order difference is a simple difference. A second-order difference is a difference in a difference. Thus, a third-order difference is a difference in a difference in a difference. In Equations 10 and 11, the INTSUM variable is used to estimate third-order differences. INTSUM sums up the three interaction variables:

$$Y = a + b_1 F_2 + b_2 F_3 + b_3 F_4 + b_4 \text{SES} + b_5 \text{INTSUM} + b_6 F_3 \text{SES} + b_7 F_4 \text{SES} \qquad (10)$$

$$\text{INTSUM} = F_2 \text{SES} + F_3 \text{SES} + F_4 \text{SES} \qquad (11)$$

U	F_2	F_3	F_4	SES	INTSUM	F_3SES	F_4SES
1	0	0	0	1	0	0	0
1	0	0	0	2	0	0	0
1	0	0	0	3	0	0	0
1	0	0	0	4	0	0	0
1	1	0	0	1	1	0	0
1	1	0	0	2	2	0	0
1	1	0	0	3	3	0	0
1	1	0	0	4	4	0	0
1	0	1	0	1	1	1	0
1	0	1	0	2	2	2	0
1	0	1	0	3	3	3	0
1	0	1	0	4	4	4	0
1	0	0	1	1	1	0	1
1	0	0	1	2	2	0	2
1	0	0	1	3	3	0	3
1	0	0	1	4	4	0	4

I replace F_2SES in Equation 9 with INTSUM in Equation 10. The coefficients for F_2SES, F_3SES, and F_4SES in Equation 9 estimated interaction effects. However, the F_3SES and F_4SES variables in Equations 10 and 11 are nested in the INTSUM variable. Thus, each coefficient now estimates the difference in the interaction effect for that variable from the interaction effect for F_2SES.

Model 1 in Table 6.8 is a standard interaction model and estimates the interaction of SES quartile with family structure. The three family structure variables are multiplied times SES to calculate the interaction variables. In Model 2, the Bio./Step. × SES variable is replaced by the INTSUM variable, which is the sum of the three interaction variables in Model 1.

Table 6.8 Interactions Between Family Structure and Parental SES With Math Score as Dependent Variable

	Model	
	1	2
Independent Variable	*B*	*B*
Bio./Step.	.03	.03
Single	.64*	.64*
Other Fam.	−1.24*	−1.24*
SES QUARTILE	3.52*	3.52*
Bio./Step. × SES QUAR.	−.64*	—
INTSUM	—	−.64*
Single × SES QUAR.	−.51*	−.13
Other Fam. × SES QUAR.	−.11	.53
Intercept	43.23	43.23
R^2	.147	.147

*p < .05.

The coefficient for the INTSUM variable in Model 2 of Table 6.8 is the same as the coefficient for Bio./Step. × SES in Model 1. When we replace a standard interaction variable such as Bio./Step. × SES with a "summer" variable like INTSUM, the summer variable will have the same coefficient as the variable that it replaced. The coefficients that change are the variables that previously were not nested in the Bio./Step. × SES variable but are now nested in the INTSUM variable. The coefficients for Single × SES and Other × SES, which estimated interaction effects like the coefficient for Bio./Step. × SES in Model 1, now estimate differences from the interaction effect for Bio./Step. × SES in Model 2.

Although the within-group effects model and the standard interaction model are basic tools for analyzing interactions, using a variable like INTSUM is likely to be a rare occurrence for most researchers. However, the example using the INTSUM variable illustrates how a "summer" variable can estimate a third-order difference, a difference of a difference of a difference. Use of the INTSUM variable also illustrates a general property of interaction variables. That is, although we can create only one set of interaction variables with two independent variables, the number of models we can create with this set of interaction variables is varied. When someone says, "Run the interactions," the appropriate response should be "Which model?"

Three-Way Interactions

In regression modeling, researchers typically use control models, but they also often use two-way interaction models. Three-way interactions models are rare in published research. The many additional variables in a three-way interaction models make interpretation difficult. Where would a researcher even start in interpreting such a model? Model 1 in Table 6.9 has four dummy variables and is an additive model. Model 2 adds seven interaction variables and is a three-way interaction model and seems complex. Model 2 has a mixture of variables, and it is not immediately obvious how to analyze such a complex model.

| Table 6.9 | Interactions Among Race/Ethnicity, Female, and College-Graduate Parent With Math Score as Dependent Variable |
| | |

| | Model | |
| | 1 | 2 |
Independent Variable	*B*	*B*
Black	−4.54*	−3.68*
Other Race/Ethnicity	.16	.06
College-Graduate Parent	6.80*	7.17*
Female	−.13	.25
College Grad. × Female	—	−.84*
Black × College Grad.	—	−3.68*
Other × College Grad.	—	.71
Black × Female	—	−.08
Other × Female	—	−.44
Black × College Grad. × Fem.	—	2.87*
Other × College Grad. × Fem.	—	.25
Intercept	49.13	48.96
R^2	.136	.138

*$p < .05$.

The challenge of interpreting a three-way interaction model is illustrated in the models in Table 6.9. The additive model includes two dummy variables for race/ethnicity, a single dummy variable for college-graduate parent, and a single dummy variable for female. The three way-interaction model then adds seven interaction variables to the additive model. There is one two-way interaction variable for the interaction between college-graduate parent and female. There are two two-way interaction variables for the interaction between race/ethnicity and college-graduate parent. There are two two-way interactions for the interaction between race/ethnicity and female. Finally, there are two interaction variables for the three-way interaction among race/ethnicity, college-graduate parent, and female. The three-way interaction model more than doubles the number of variables in the additive model. Interpreting this model seems difficult as a result of the number and variety of variables.

In two-way interaction models, the standard interaction model estimates differences between coefficients in the within-group effects model. Thus, it is necessary to choose the within-group effects model of interest to interpret the standard interaction model properly. We can take a similar approach in working with a three-way interaction model.[5] There will also be a standard interaction model and a within-group effects model in the three-way interaction case.

Specifying the within-group model for three-way interactions involves focusing on one set of two-way interactions. Although there are three alternative two-way models in this example, the focus in Model 2 in Table 6.10 is on the two-way interaction between college-graduate parent and female. Model 3 is the same as the three-way interaction model in Table 6.9, but the variables are listed in a specific order. The three-way interaction model can be viewed as taking the two additive variables and the two-way interaction for college-graduate parent and female and then interacting those three variables with race/ethnicity.

Model 1 in Table 6.11 is the standard interaction model. At the top of the column are the variables for Black and other, which will serve as the conditioning variables. Next are variables for college-graduate parent, female, and the two-way interaction between those variables. Again, the focus is on the two-way interaction between college-graduate parent and female when interpreting the three-way interaction analysis. Model 1 also includes six additional variables that capture the three-way interaction.

Within Model 2 in Table 6.11 are standard interaction models for the interaction between college-graduate parent and female for Whites, Blacks, and other, separately. Thus, Model 1 can be viewed as testing whether the interactions between college-graduate parent and female are different for Blacks and other as compared with Whites. Model 2, on the other hand, shows the two-way interaction model for each race/ethnic group.

---

5. Jaccard (2001), pages 24–30, suggests that one approach to making three-way interactions more interpretable is to view three-way interactions as two-way interactions conditioned on a third variable.

Table 6.10 Interactions Among Race/Ethnicity, Female, and College-Graduate Parent With Math Score as Dependent Variable

	Model		
	1	**2**	**3**
Independent Variable	**B**	**B**	**B**
Black	−4.54*	−4.54*	−3.68*
Other Race/Ethnicity	.16	.17	.06
College-Graduate Parent	6.80*	7.04*	7.17*
Female	−.13	.08	.25
College Grad. × Female	—	−.48	−.84*
Black × College Grad.	—	—	−3.68*
Black × Female	—	—	−.08
Black × College Grad. × Fem.	—	—	2.87*
Other × College Grad.	—	—	.71
Other × Female	—	—	−.44
Other × College Grad. × Fem.	—	—	.25
Intercept	49.13	49.03	48.96
R^2	.136	.136	.138

*p < .05.

The key to a three-way interaction analysis is whether the three-way interaction coefficient is significantly different from zero. In this particular analysis, the key variable is the Black × college-graduate parent × female interaction. The three-way interaction involving other race/ethnicity is not substantively interesting.

The coefficient for the Black × college-graduate parent × female variable is significant, and interpretation of the three-way interaction model is appropriate. Examining the within-group two-way interactions in Model 2 in Table 6.11 helps greatly in interpreting the three-way interaction. The three-way interaction coefficient in Model 1 is 2.87. This is the difference between the coefficient two-way interaction between college-graduate parent and female for Blacks and the coefficient for Whites (2.03 − (−.84)). These two-way interaction coefficients can be interpreted in two ways. I interpret them as the additional female effect for those with a college-graduate parent.

Table 6.11 **Interactions Among Race/Ethnicity, Female, and College-Graduate Parent With Math Score as Dependent Variable**

Independent Variable	Model				
	1	**2**	**3**	**4**	**5**
	B	**B**	**B**	**B**	**B**
Black	−3.68*	−3.68*	—	—	—
Other Race/Ethnicity	.06	.06	—	—	—
College-Graduate Parent	7.17*	—	7.17*	3.49*	7.88*
Female	.25	—	.25	.17	−.19
College Grad. × Female	−.84*	—	−.84*	2.03*	−.60
White × College Grad.	—	7.17*	—	—	—
White × Female	—	.25	—	—	—
White × College Grad. × Fem.	—	−.84*	—	—	—
Black × College Grad.	−3.68*	3.49*	—	—	—
Black × Female	−.08	.17	—	—	—
Black × College Grad. × Fem.	2.87*	2.03*	—	—	—
Other × College Grad.	.71	7.88*	—	—	—
Other × Female	−.44	−.19	—	—	—
Other × College Grad. × Fem.	.24	−.60	—	—	—
Intercept	48.96	48.96	48.96	45.28	49.02
R^2	.138	.138	.117	.063	.126

*$p < .05$.

Also, the last three columns in Table 6.11 show models for the two-way interaction between college-graduate parent and female for Whites only in Model 3, for Blacks only in Model 4, and for other race/ethnicity only in Model 5. The coefficients in these three models are the same as the coefficients in the within-group model, Model 2. Thus, the standard interaction model, Model 1, tests whether the respective coefficients in Models 3, 4, and 5 are equal. This means that running separate interaction models is equivalent to running interaction models with the race/ethnicity variable as the conditioning variable.

Table 6.12 Interactions Among Race/Ethnicity, Female, and College-Graduate Parent With Math Score as Dependent Variable

	Model		
	1	2	3
Independent Variable	**B**	**B**	**B**
Black	−3.68*	−3.68*	−3.68*
Other Race/Ethnicity	.06	.06	.06
College-Graduate Parent	7.17*	—	—
Female	.25	—	—
College Grad. × Female	−.84*	—	—
White × College Grad.	—	7.17*	7.17*
White × Female	—	.25	—
White × Not College Grad. × Fem.	—	—	.25
White × College Grad. × Fem.	—	−.84*	−.59*
Black × College Grad.	−3.68*	3.49*	3.49*
Black × Female	−.08	.17	—
Black × Not College Grad. × Fem.	—	—	.17
Black × College Grad. × Fem.	2.87*	2.03*	2.20*
Other × College Grad.	.71	7.88*	7.88*
Other × Female	−.44	−.19	—
Other × Not College Grad. × Fem.	—	—	−.19
Other × College Grad. × Fem.	.24	−.60	−.79*
Intercept	48.96	48.96	48.96
R^2	.138	.138	.138

*$p < .05$.

In the discussion on two-way interactions, I suggested that the best way to interpret a two-way interaction is to interpret the within-groups model. Table 6.12 shows a further developed within-groups model for the three-way interaction. In this model, there is a separate female effect for those not with a college-graduate parent and for those with a college-graduate parent within each race/ethnic group. The female effect for Whites not with a college graduate parent is not significant, whereas the female effect for Whites with a college-graduate parent is significantly negative. In contrast, the female effect for Blacks not with a college-graduate parent is not significant, whereas the female effect for Blacks with a college-graduate parent is significantly positive.

White females with a college-graduate parent score lower in math than males do, but the reverse is true for Blacks. Black females with a college-graduate parent score higher in math than do males. A college-graduate parent disadvantages females among Whites but advantages females among Blacks. In both groups, there is no sex difference for those not with a college-graduate parent.

Estimating Separate Models

In my analysis of three-way interactions, I showed that we can view a three-way inter-action model as a set of two-way interaction models conditioned on a third variable. We can use this modeling approach to test whether coefficients for subgroups in sepa-rate models are equal. For example, a researcher may use regression modeling to arrive at a specific model and then look to see whether the coefficients in the model are dif-ferent for Whites, Blacks, and other race/ethnicity.[6] One approach to determining whether the models are different from one another is the Chow test.[7] However, we can conduct an equivalent test that uses interactions.

Table 6.13 shows separate models for Whites, Blacks, and other race ethnicity for the effect of family structure, college-graduate parent, and family income on math scores.

The results indicate that although there is a significant positive effect of two-biological-parent family for White, there is no significant effect for Blacks. In addition, the effect for college-graduate parent is less for Blacks than Whites, but the effect of income is larger for Blacks than for Whites.

The method of using interactions to estimate separate models interacts all the independent variables with the conditioning variable.[8] In this case, race/ethnicity is the conditioning variable and we interact each of the three dummy variables representing race/ethnicity with the other independent variables.

Table 6.14 shows the results of regression involving interacting race/ethnicity with the other independent variables. Model 1 is the additive model, and Model 2 shows the

---

6. Hardy (1993), page 49, and Jaccard (2001), page 17, make the important point that the prob-lem with estimating separate models and comparing coefficients across models is that no significance test for the difference between coefficients is conducted.

7. Demaris (2004), pages 110–112, and Gordon (2010), pages 277–286, present the Chow test for testing the difference between separate models. I feel using the F test for linear regression or the chi-square test for logistic regression for testing the difference in fit between the additive model and the with-in group effects model is a simpler approach.

8. Demaris (2004), pages 151–152, and Hill, Griffiths, and Judge (1997), pages 190–192, discuss how to use interactions to do a global test for the difference between similar regression models for two or more groups.

Table 6.13 Separate Models for Whites, Blacks, and Other Race/Ethnicity for Effects of Family Structure, College-Graduate Parent, and Family Income With Math Score as Dependent Variable

Independent Variable	White B	Black B	Other B
Two-Biological-Parent Family	1.15*	.81	1.39*
College-Graduate Parent	5.48*	3.31*	6.15*
Family Income	.14*	.17*	.17*
Intercept	47.61	44.30	47.36
R^2	.141	.080	.151

*p < .05.

standard interaction model that allows the effects of two-biological-parent family, college-graduate parent, and family income to vary by race/ethnicity. When we focus on the White/Black contrast, we observe that Model 2 indicates that the difference in the effects of two-biological-parent family and family income are not significantly different from one another. However, the coefficient for the Black × college-graduate parent is significant, indicating that the effect of college-graduate parent is less for Blacks than for Whites.

Model 3 in Table 6.14 is the within-group effects interaction model. Notice that the coefficients for two-biological-parent family, college-graduate family, and family income for Whites, Blacks, and other race/ethnicity are the same as those for the separate models in Table 6.13. Model 4 is what I call the "separate models model." This model is estimated by suppressing the intercept and adding a dummy variable for White to the model. The coefficients for two-biological-parent family, college-graduate parent, and family income are the same as in Model 3. The difference between Model 3 and Model 4 is that the coefficients for Black and other race/ethnicity in Model 4 are no longer additive variables as in Model 3 but are now the same as the intercepts for the separate models as in Table 6.12.

I do not recommend estimating the separate models model except for the purpose of learning and exploring regression modeling. However, the separate models model does show that it is possible to estimate one regression model that replicates exactly the results from running separate models.

Example Using Logistic Regression

Family income has an important influence on whether a student attends a private high school. Those with families with higher incomes are more likely to attend since private high schools often cost substantially more than public schools.

| Table 6.14 | Interactions Among Race/Ethnicity and Family Structure, College-Graduate Parent, and Family Income With Math Score as Dependent Variable |

	Model			
	1	**2**	**3**	**4**
Independent Variable	**B**	**B**	**B**	**B**
White	—	—	—	47.61
Black	−3.98*	−3.31*	−3.31*	44.30
Other Race/Ethnicity	.41*	−.26	−.26	47.36
Two-Biological-Parent Family	1.19*	1.15*	—	—
College-Graduate Parent	5.49*	5.48*	—	—
Family Income	.15*	.14*	—	—
White × Two-Parent Family	—	—	1.15*	1.15*
White × College-Graduate Parent	—	—	5.48*	5.48*
White × Family Income	—	—	.14*	.14*
Black × Two-Parent Family	—	−.34	.81	.81
Black × College-Graduate Parent	—	−2.17*	3.31*	3.31*
Black × Family Income	—	.03	.17*	.17*
Other × Two-Parent Family	—	.24	1.39*	1.39*
Other × College-Graduate Parent	—	.67*	6.15*	6.15*
Other × Family Income	—	.03*	.17*	.17*
Intercept	47.46	47.61	47.61	—
R^2	.160	.162	.162	—

*$p < .05$.

The analysis in Table 6.15 considers whether the effect of family income ($10,000 units) is the same for Blacks as for Whites. Model 1 shows a significant effect of income in the additive model. Model 2 is the standard interaction model and shows a significantly higher effect for family income for Blacks than for Whites. The coefficient for family income in Model 3 is .08 for Blacks and .06 for Whites. Each additional unit of family income has a greater impact on Black chances than on those of Whites.

Table 6.15   Interactions Among Race/Ethnicity and Family Income With Private High School as Dependent Variable

	Model		
	1	2	3
Independent Variable	B	B	B
Black	−.13	−.30*	−.30*
Other Race/Ethnicity	−.21*	−.22*	−.22*
Family Income	.06*	.06*	—
White × Fam. Inc.	—	—	.06*
Black × Fam. Inc.	—	.02*	.08*
Other × Fam. Inc.	—	.00	.06*
Intercept	−2.15	−2.14	−2.14
−2 log-likelihood	16,639.7	16,632.0	16,632.0

*p < .05.

The analysis in Table 6.16 considers the interaction between race/ethnicity and whether the parent is a college graduate on chances of attending private high school. When the variables involved in the interaction are both categorical, the researcher has a choice about which variable to condition on the other. In the analysis in Table 6.16, college-graduate parent is conditioned on race/ethnicity.

The coefficient for Black × College Grad. is significant and negative. This indicates that the effect of college-graduate parent is less for Blacks than for Whites. Model 3 in Table 6.16 shows that the effect of college-graduate parent is 1.46 for Whites and 1.18 for Blacks. Having a college-graduate parent is less of an advantage for Blacks than for Whites.

In Table 6.17, race/ethnicity is conditioned on college-graduate parent. The coefficient for Black × College Grad. in Model 2 is significant and negative. This indicates that the Black coefficient is more negative among those who have a college-graduate parent than among those who do not have a college-graduate parent. Model 3 shows that the Black coefficient among those who do not have a college-graduate parent is −.03 and not significant and that the coefficient for those with a college-graduate parent is −.31 and significant. There is a racial difference in chances of attending private high school among those who have a college-graduate parent but not among those who do not have a college-graduate parent.

Table 6.16 Interactions Between Race/Ethnicity and College-Graduate Parent With Private High School as Dependent Variable

	Model		
	1	**2**	**3**
Independent Variable	*B*	*B*	*B*
Black	−.19*	−.03	−.03
Other Race/Ethnicity	−.22*	−.15*	−.15*
College-Graduate Parent	1.40*	1.46*	—
White × College Grad.	—	—	1.46*
Black × College Grad.	—	−.28*	1.18*
Other × College Grad.	—	−.11	1.35*
Intercept	−2.30	−2.34	−2.34
−2 log-likelihood	16,757.1	16,752.4	16,752.4

*p < .05.

Table 6.17 Interactions Between Race/Ethnicity and College-Graduate Parent With Private High School as Dependent Variable

	Model		
	1	**2**	**3**
Independent Variable	*B*	*B*	*B*
College-Graduate Parent	1.40*	1.46*	1.46*
Black	−.19*	−.03	—
Other Race/Ethnicity	−.22*	−.15*	—
Not College Grad. × Black	—	—	−.03
Not College Grad. × Other	—	—	−.15*
College Grad. × Black	—	−.28*	−.31*
College Grad. × Other	—	−.11	−.26*
Intercept	−2.30	−2.34	−2.34
−2 log-likelihood	16,757.0	16,752.4	16,752.4

*p < .05.

Summary

An additive model constrains the effect of an independent variable to be the same in categories of a second independent variable. Estimating interaction models allows the researcher to remove this constraint and to allow the effects of an independent variable to vary for subgroups.

I argue that for any standard interaction model, there is a within-group effects model that serves as the basis for the standard interaction model. The within-group effects interaction model estimates the effect of an independent variable within subgroups defined by a second independent variable. The standard interaction model then provides the test for whether the effect of an independent variable is equal across subgroups. I suggest that a common reason that the standard interaction model can be misinterpreted or can be difficult to interpret is a result of the lack of a clearly defined and recognized within-group effects interaction model.

This chapter does not address interactions between two interval variables. I find that interactions between two interval variables are difficult to interpret. Since the coefficient is not readily interpretable, a fuller understanding of the meaning of the interaction coefficient for two interval variables is best achieved by graphing the relationship measured by the interaction coefficient.[9] In contrast, I show in this chapter that interactions involving independent variables represented by dummy variables are readily interpretable. Therefore, I suggest that a more interpretable way to analyze interactions between two interval variables is to convert one of the interval variables into a categorical variable captured by dummy variables. I would suggest at least three or four categories for the recoded interval variable to model the effects of the variable adequately. Although such an approach is less parsimonious because a variable whose effect on the dependent variable was captured with one coefficient is now captured with multiple coefficients, the advantage of interpretability outweighs, in my view, any loss of parsimony.

When doing regression analysis using interaction variables, the researcher must pay close attention to the sample size in each sample subgroup. One rule of thumb is that if a sample size of 25 is needed to get a good estimate of a population mean, then any subgroups as defined by an interaction variable need to have sample sizes of at least 25. So after creating a dummy variable that represents an interaction, look at the frequency distribution for the variable to make sure that both of the categories defined by the dummy variable have a sample size of at least 25.

Key Concepts

conditional differences: the idea that the difference between two groups on a dependent variable will vary within categories of a third variable.

two-way interaction: a variable constructed by multiplying two independent variables; when one of the variables involved is a dummy variable, then the interaction variable will be a subset of one or both of the independent variables involved in the interaction.

---

9. Gordon (2010) provides a brief treatment of interactions between interval variables. Jaccard (1990, 2001) provide more extensive treatments.

standard interaction model: an interaction model that incorporates interactions between two sets of independent variables; the interaction coefficients in a standard interaction model measure the additional effect of one of the independent variables for one group compared with the effect of the same independent variable for another group when at least one of the sets of independent variables is measured with dummy variables.

within-group effects interaction model: an interaction model that incorporates interactions between two sets of independent variables; the interaction coefficients in a within-group effects interaction model measure the effect of one of the independent variables for one group and the effect of that same independent variable for another group when at least one of the sets of independent variables is measured with dummy variables.

three-way interactions: a variable constructed by multiplying three independent variables; to enhance interpretability, I suggest that researchers view a three-way interaction as a set of two-way interactions conditioned on a third independent variable.

separate models interaction model: a two-way interaction model that estimates in one model separate regression models for subgroups defined by a third independent variable; the objective of the discussion of the separate models model was to show that interactions can be used to estimate the same coefficients that can be obtained by estimating separate regressions for subgroups.

Chapter Exercises

1. Replicate the regressions and the table for the interaction between race/ethnicity and college-graduate parent with private as the dependent variable like in Table 6.14. Use PRIVATE, BLACK, OTHRACE, PARCOLL, WPARCOLL, BPARCOLL, and OPARCOLL in the analysis.

2. Run regressions using the linear regression procedure, and create a table similar to the one for the interaction between race/ethnicity and college-graduate parent with math score as the dependent variable. Use X2TXMTSCOR, BLACK, OTHRACE, PARCOLL, WPARCOLL, BPARCOLL, and OPARCOLL in the analysis.

   How do the effects of college-graduate parent compare for Whites and Blacks?

3. Run regressions using the linear regression procedure, and create a table for the interaction between family structure and family income with math score as the dependent variable like in Table 6.7. You will use the following variables in creating the models: two biological parent, biological parent/stepparent, single, other family, and family income. Use X2TXMTSCOR, STEP, SINGLE, FAMOTH, FAMINC, TWOINC, STEPINC, SINGLEINC, and FAMOTHINC in the analysis.

   How do the effects of family income differ for those in two-biological-parent, biological-parent/stepparent, and single-parent families?

7

Modeling Linearity With Splines

Dummy Variables Nested in Interval Variable

It may seem that using dummy variables to model an independent variable in regression modeling and using an interval variable to model an independent variable are unrelated approaches.

In the discussion that follows, however, first I show how in a special case, the interval variable model can be nested within the dummy variable model. Second, I show how using the same logic as in the special case any interval variable model can be nested within a less constrained spline variable model.

First, I provide an example using general notation. Suppose we have an interval variable that has four values, 1, 2, 3, and 4. The interval variable is shown as follows along with the unit vector:

$$
\begin{array}{cc}
U & INTERVAL \\
\begin{bmatrix} 1 \\ 1 \\ 1 \\ 1 \end{bmatrix} &
\begin{bmatrix} 1 \\ 2 \\ 3 \\ 4 \end{bmatrix}
\end{array}
$$

We could also express the information in the data with four dummy variables. In this case, there are four groups and four means, so the dummy variable regression model will capture all available information. The objective of this example is to show how the interval variable regression model places constraints on this basic information.

It is possible to enter all four dummy variables shown as follows into a regression model. This would require that the intercept be suppressed. The resulting coefficients would estimate the mean for each group if ordinary least-squares (OLS) regression was used:

$$
\begin{array}{cccc}
D_1 & D_2 & D_3 & D_4 \\
\begin{bmatrix} 1 \\ 0 \\ 0 \\ 0 \end{bmatrix} &
\begin{bmatrix} 0 \\ 1 \\ 0 \\ 0 \end{bmatrix} &
\begin{bmatrix} 0 \\ 0 \\ 1 \\ 0 \end{bmatrix} &
\begin{bmatrix} 0 \\ 0 \\ 0 \\ 1 \end{bmatrix}
\end{array}
$$

The variables that we would use in the typical dummy variable regression are shown as follows. D_1 is not included, and U is included. U is the sum of $D_1+D_2+D_3+D_4$. U would estimate the mean for group 1, and the coefficients for D_2, D_3, and D_4 would estimate differences from that mean:

$$
\begin{array}{cccc}
U & D_2 & D_3 & D_4 \\
\begin{bmatrix} 1 \\ 1 \\ 1 \\ 1 \end{bmatrix} &
\begin{bmatrix} 0 \\ 1 \\ 0 \\ 0 \end{bmatrix} &
\begin{bmatrix} 0 \\ 0 \\ 1 \\ 0 \end{bmatrix} &
\begin{bmatrix} 0 \\ 0 \\ 0 \\ 1 \end{bmatrix}
\end{array}
$$

We can use nested variables to estimate the difference between the coefficients for D_3 and D_4. We do this by creating a larger variable within which the smaller variable will be nested. I name the new variable J_3.

The new variable J_3 is the sum of D_3 and D_4. I call the new variable J_3 because when the variable is entered into the model, it causes the D_4 coefficient in the new model to estimate the difference between, or jump from, the D_3 coefficient in the previous model to the D_4 coefficient in that model. The coefficient for the J_3 variable in the new model will be the same as the D_3 coefficient in the previous model:

$$J_3 = D_3 + D_4$$

$$
\begin{array}{cccc}
U & D_2 & J_3 & D_4 \\
\begin{bmatrix} 1 \\ 1 \\ 1 \\ 1 \end{bmatrix} &
\begin{bmatrix} 0 \\ 1 \\ 0 \\ 0 \end{bmatrix} &
\begin{bmatrix} 0 \\ 0 \\ 1 \\ 1 \end{bmatrix} &
\begin{bmatrix} 0 \\ 0 \\ 0 \\ 1 \end{bmatrix}
\end{array}
$$

In the same manner, I can create another new variable, J_2, which is the sum of D_2 and J_3. The coefficient for this new variable will estimate the difference or jump between the D_2 variable and the J_3 variable. The result is a model that estimates the mean for the first group, the difference between the means for the first and the second groups, the difference between the second and third groups, and the difference between the third and fourth groups. The differences can be viewed as jumps from one group to another so I refer to this kind of model as a "jump" model:

$$J_2 = D_2 + J_3$$

$$
\begin{array}{cccc}
U & J_2 & J_3 & D_4 \\
\begin{bmatrix} 1 \\ 1 \\ 1 \\ 1 \end{bmatrix} &
\begin{bmatrix} 0 \\ 1 \\ 1 \\ 1 \end{bmatrix} &
\begin{bmatrix} 0 \\ 0 \\ 1 \\ 1 \end{bmatrix} &
\begin{bmatrix} 0 \\ 0 \\ 0 \\ 1 \end{bmatrix}
\end{array}
$$

The basic assumption of linearity is that for each increment of change in the independent variable, the dependent variable changes the same amount each time. For the previous model, this assumption would require that the jumps be equal. We can test this assumption by summing J_2, J_3, and D_4 to create S_1. By including S_1 in the model rather than J_2, the coefficients for J_3 and D_4 in the new model measure the difference between the coefficients for J_2 and J_3 and between J_2 and D_4 in the previous model. Note that S_1 looks like an interval variable:

$$S_1 = J_2 + J_3 + D_4$$

$$
\begin{array}{cccc}
\text{U} & S_1 & J_3 & D_4 \\
\begin{bmatrix} 1 \\ 1 \\ 1 \\ 1 \end{bmatrix} &
\begin{bmatrix} 0 \\ 1 \\ 2 \\ 3 \end{bmatrix} &
\begin{bmatrix} 0 \\ 0 \\ 1 \\ 1 \end{bmatrix} &
\begin{bmatrix} 0 \\ 0 \\ 0 \\ 1 \end{bmatrix}
\end{array}
$$

The final step is to constrain the jumps to be equal. We can set the coefficients for J_2 and D_4 to be zero by not including the variables in the model:

$$
\begin{array}{cc}
\text{U} & S_1 \\
\begin{bmatrix} 1 \\ 1 \\ 1 \\ 1 \end{bmatrix} &
\begin{bmatrix} 0 \\ 1 \\ 2 \\ 3 \end{bmatrix}
\end{array}
$$

The constrained model looks very much like the interval model that we started with. In fact, S_1 is a transformation of the interval variable since INTERVAL minus 1 equals S_1:

$$S_1 = \text{INTERVAL} - 1$$

We can substitute INTERVAL for S_1 in the model, and the coefficient for INTERVAL will be the same as the coefficient for S_1 as will the R^2 because INTERVAL is a linear transformation of S_1. This completes my proof that in this special case, the model with the INTERVAL variable is nested within the model with the three dummy variables, D_2, D_3, and D_4. This example illustrates the principle that if we constrain coefficients to be equal to zero, the smaller model is nested within the larger model. The example has illustrated the key concepts of nestedness and constraints. It also has illustrated spline variables without directly saying so:

$$
\begin{array}{cc}
\text{U} & \text{INTERVAL} \\
\begin{bmatrix} 1 \\ 1 \\ 1 \\ 1 \end{bmatrix} &
\begin{bmatrix} 1 \\ 2 \\ 3 \\ 4 \end{bmatrix}
\end{array}
$$

Table 7.1 shows an empirical example to illustrate the relationship between an integer interval variable and a dummy variable model that was explained earlier. The interval variable in this case is socioeconomic status (SES) quartile and has values 1, 2, 3, and 4. The coefficients for the second quartile, third quartile, and fourth quartile dummy variables measure differences from the first quartile mean as captured by the intercept.

The second model includes the third quartile jump variable in place of the third quartile dummy variable. The third quartile jump variable is the sum of the third quartile dummy variable and the fourth quartile dummy variable. The fourth quartile dummy variable is nested in the third quartile jump variable.

The fourth quartile dummy coefficient in Model 2 is the difference between the third and fourth quartile dummy coefficients in Model 1. In Model 3, I substitute the second quartile jump variable for the second quartile dummy variable. The second quartile jump variable is the sum of the second quartile dummy variable and the third quartile jump variable. The third quartile jump variable in Model 3 is nested in the second quartile jump variable.

The results in Model 3 show how much more the second quartile mean is than the first quartile mean (second quartile jump coefficient), how much more the third quartile mean is than the second quartile mean (third quartile jump coefficient), and how much more the fourth quartile mean is than the third quartile mean (fourth quartile

Table 7.1 Using Dummy and Jump Variables to Model the Effect of SES Quartile With Math Score as Dependent Variable

	Model			
	1	2	3	4
Independent Variable	**B**	**B**	**B**	**B**
Second Q Dummy	2.72*	2.72*	—	—
Second Q Jump	—	—	2.72*	—
Jump Summer	—	—	—	2.72*
Third Q Dummy	5.31*	—	—	—
Third Q Jump	—	5.31*	2.59*	−.13
Fourth Q Dummy	10.54*	5.23*	5.23*	2.51*
Intercept	46.97	46.97	46.97	46.97
R^2	.148	.148	.148	.148

*$p < .05$.

dummy coefficient). The fourth quartile dummy coefficient is larger than the second and third quartile coefficients, which is not consistent with the linear assumption.

We can test for equality in these coefficients by substituting the "jump summer" for the second quartile jump variable. This variable is the sum of the second and third quartile jump variables and the fourth quartile dummy variable. The third quartile jump variable and the fourth quartile dump variables are nested in the jump summer variable. This model tests whether the jumps from the second quartile to the third quartile and from third to the fourth are equal to the jump from the first to the second. The coefficient for the third quartile jump variable in Model 4 is not significant, whereas the coefficient for the fourth quartile dummy variable is significant so the results are mixed. The significant coefficient is evidence that the relationship between SES quartile and math score is not perfectly linear.

Model 1 in Table 7.2 is the dummy variable model, and Model 2 is the model that includes the jump summer variable. Model 3 constrains the third quartile jump coefficient and the fourth quartile dummy coefficient to be zero. This is done by estimating the regression model without including these variables. Model 3 is nested in Model 2.

The jump summer coefficient in Model 3 in Table 7.2 is 3.43 and is larger than the jump summer coefficient in the unconstrained Model 2. In Model 4, I include only the original interval SES quartile variable, and the coefficient for that variable is also 3.43.

Table 7.2 Using Dummy and Jump Variables to Model Effect of SES Quartile With Math Score as Dependent Variable

	Model			
	1	2	3	4
Independent Variable	*B*	*B*	*B*	*B*
Second Q Dummy	—	—	—	—
Second Q Jump	2.72*	—	—	—
Jump Summer	—	2.72*	3.43*	—
Interval Q Variable	—	—	—	3.43*
Third Q Dummy	—	—	—	—
Third Q Jump	2.59*	−.13	—	—
Fourth Q Dummy	5.23*	2.51*	—	—
Intercept	46.97	46.97	46.48	43.05
R^2	.148	.148	.143	.143

*$p < .05$.

The only difference between Model 3 and Model 4 in Table 7.2 is in the intercepts. The intercept in Model 3 is the estimated mean for the first SES quartile under the constraint of linearity. The intercept in Model 4 is the estimated mean for the group, which is zero on SES quartile under the linear constraint. Since the SES variable is coded 1, 2, 3, and 4, there is no actual group with a zero value. Note that if we add 3.43 to 43.05, the result is 46.48.

Thus, if we wanted the intercept to be meaningful in regression models that include interval variables, we could transform all interval variables such that the new variables were deviations from the lowest value. The result would be that the lowest value would have a score of zero and the intercept in a regression model that used the transformed interval variables would be meaningful. However, since we rarely interpret the intercept in creating research results, there really is no point in transforming the interval variables in this manner. A linear transformation such as in Table 7.2 only changes the intercept and not the other coefficients.

Introduction to Knotted Spline Variables

When an interval variable has too many values for the variable to be captured by dummy variables, an alternative approach to assessing the linear relationship is to use knotted spline variables.[1] The method of knotted splines allows us to visualize the relationship between the independent and dependent variables as a series of connected line segments with differing slopes rather than by a straight line with one slope.[2] Using spline variables to model a linear relationship may allow the researcher to create an equation that fits the data better than the equation that includes only one linear coefficient. Thus, the reason a researcher will explore using spline variables is to find a better fitting model.

The procedure involves creating variables that measure how much of a particular segment of an interval variable is captured for each unit of analysis. For example, consider an interval variable for number of siblings that has values 0 through 8. Suppose we divide the number of siblings variable into three segments: 0–2, 3–5, and 6–8. We can easily think of these segments as capturing for each respondent the number of siblings with birth order 0 through 2, the number with birth order 3 through 5, and the number with birth order 5 through 8.

---

1. Hosmer, Lemeshow, and Sturdivant (2013), pages 99–100, and Greene (2012), Chapter 6, discuss knotted splines and a two-variable difference spline model. Hardy (1993), pages 80–82, discusses piecewise linear regression, which is the same as knotted spline regression. Hardy (1993) shows the difference spline model.

2. The use of splines is a less commonly used method for examining nonlinearity. However, I argue that using splines is a more interpretable method than using mathematical transformations. Gordon (2010), Chapter 9, presents the standard approach by using mathematical transformations. See also Demaris (2004), Chapter 5, and Agresti and Finlay (2009), pages 462–473. Allison (1999), Chapter 8, and Linneman (2014), Chapter 15, provide basic introductions to modeling linearity by using squared terms and by transforming the dependent variable.

The advantage to using knotted splines to model a variable is that there may be differing returns to units. The influence of having a sibling on the dependent variable for the first two siblings may be different than the influence for the third through fifth siblings or the sixth through eighth siblings. The following table shows how the values of the spline variables would look for various values of the number of siblings variable.

For each respondent, there is a value for each spline variable depending on the number of siblings for the respondent. Those with zero siblings are zero on all the spline variables. Those with three siblings are two on the 0–2 spline variable because they have a first and a second birth order sibling and 1 on the 3–5 spline variable because they have a third birth order sibling. Those with six siblings have two siblings who are birth order 0–2, three siblings who are birth order 3–5, and one sibling who is birth order 6–8.

Notice that for each value for the number of siblings, the three spline variables sum up to that number. It is like we pulled the interval variable apart and split it into three subparts. The implication is that a model with the original interval variable is a constrained version of the model that would include the three spline variables:

| | Spline variables | | |
Siblings	0–2	3–5	6–8
0	0	0	0
1	1	0	0
2	2	0	0
3	2	1	0
4	2	2	0
5	2	3	0
6	2	3	1
7	2	3	2
8	2	3	3

Spline Variables Nested in an Interval Variable

The example using the number of siblings variable has been useful in explaining how spline variables are constructed. However, splines are also useful for modeling interval variables that have even more values. In the next example, I will create spline variables for a modified SES variable that has 12 values. To have a manageable number of values for this example, I transformed the original SES variable so that it had 12 values from 1 to 12. The 12 values evenly cover the range of the original SES variable. The number of spline variables that the researcher creates will depend on the overall sample size since there should be at least 25 to 50 units of analysis in each segment. Generally,

I suggest that researchers start with between 4 and 12 spline variables to keep the analysis perceptually manageable.

In the following table, I show six spline variables. Each variable captures a unique segment of the range for the revised SES variable. Altogether, the five variables capture all the range of the revised SES variable. The spline variables must cover all the range the original variable and not overlap.

The variable S_1_2 captures all the values of the revised SES variable that occurs in the range between 1 and 2. Those respondents who had a score of 1 on the revised SES variable, score 1 on S_1_2, whereas those respondents who had a score of 2 on the revised SES variable score 2 on S_1_2. However, those respondents who had a score of 3 on the SES variable, score 2 on the S_1_2, and that is the maximum that any respondent can score on this variable. Note that if the five spline variables are added together, the sum is the same as the revised SES variable:

SES-revised	S_1_2	S_3_4	S_5_6	S_7_8	S_9_10	S_11_12
1	1	0	0	0	0	0
2	2	0	0	0	0	0
3	2	1	0	0	0	0
4	2	2	0	0	0	0
5	2	2	1	0	0	0
6	2	2	2	0	0	0
7	2	2	2	1	0	0
8	2	2	2	2	0	0
9	2	2	2	2	1	0
10	2	2	2	2	2	0
11	2	2	2	2	2	1
12	2	2	2	2	2	2

The variable S_9_10 captures all the value of the revised SES variable that occurs in the range between 9 and 10. For the S_9_10 variable, those respondents who had a score of 8 on the revised SES variable score zero on this variable because no part of their score falls in the 9–10 interval. Those respondents who had a score of 9 on the revised SES variable score 1 on S_9_10. Finally, those respondents who had a score of 10 on the revised SES variable score 2 on S_9_10, and that is the maximum score on this variable. In the following table, the ranges covered by each variable are listed. A respondent may have a nonzero value for one, some, or all of the variables (see the table on the top of the next page).

The regression models in Table 7.3 include SES and the SES knotted splines as independent variables and math score as the dependent variable. Another way to think about

S_1_2	$1 \leq$ number of SES-revised units ≤ 2
S_3_4	$3 \leq$ number of SES-revised units ≤ 4
S_5_6	$5 \leq$ number of SES-revised units ≤ 6
S_7_8	$7 \leq$ number of SES-revised units ≤ 8
S_9_10	$9 \leq$ number of SES-revised units ≤ 10
S_11_12	$11 \leq$ number of SES-revised units ≤ 12

the coefficient for the revised SES variable is that it is the coefficient that results if the coefficients for the knotted spline variables are constrained to be equal. The implication is that the revised SES coefficient will fall somewhere within the range of coefficients for the knotted spline variables. The results show that the coefficient for revised SES is significantly different from zero as are all the coefficients for the knotted spline variables. The coefficients for the knotted spline variables generally increase in size going from the S_1_2 to S_11_12. The slopes for the knotted spline variables are all positive but vary in size.

Table 7.3   Using Spline Variables to Model the Effect of the Revised SES Variable With Math Score as Dependent Variable

	Model	
	1	**2**
Independent Variable	***B***	***B***
SES-revised	1.78*	—
S_1_2	—	1.18*
S_3_4	—	1.39*
S_5_6	—	1.31*
S_7_8	—	2.13*
S_9_10	—	2.24*
S_11_12	—	2.07*
Intercept	40.90	43.07
R^2	.158	.161

*$p < .05$.

We can use an F test to decide statistically whether the model with the spline variables is significantly better than the model with the SES variable.[3] An F test checks whether the increase in R^2 is significantly large enough to justify the addition of the extra coefficients used in the spline model:

$$F = \frac{\Delta R^2 / r}{(1 - R^2) / (n - k - 1)}$$

where

F = F statistic

$\Delta R^2 = R^2$ (big model) $- R^2$ (small model)

r = additional coefficients in the big model

$R^2 = R^2$ for the big model

n = sample size

k = number of coefficients in the big model

The F statistic calculates the increase in R^2 per additional coefficient and compares that with the remaining unexplained variation per unused degree of freedom. A significant F statistic is one where the increase in R^2 per coefficient is large relative to average unexplained variation. The following F statistic tests for improvement in R^2 due to using the spline variables (R^2 is .158208 for the small model and .161125 for the big model; sample size is 20,133):

$$F = \frac{(.161125 - .158208) / 5}{(1 - .161125) / (20{,}133 - 6 - 1)} = 14.00$$

The critical F statistic at the .05 level with (5, ∞) degrees of freedom for the numerator and denominator is 2.21. Since the calculated F statistic of 14.00 is greater than the critical F statistic of 2.21, using knotted spline variables led to a significant improvement in R^2 over the model that used the revised SES variable alone. The variation in the size of the coefficients for the knotted spline variables was enough to lead to a significant result.

I used the letter "S" on the knotted spline variables to denote spline, but we can also view the letter as standing for "segment spline." A segment spline variable is one that captures a unique segment of the range of the interval variable.[4] Once we have estimated the segment spline variables, the next question to answer is, "Are they different from one another?" We can answer this question by using the principle of nesting.

3. Allen (1997), pages 113–117, provides a discussion of nested models and how to use the F test to test changes in model fit between nested models. See Agresti and Finlay (2009), page 337, for how to calculate an F test.

4. Treiman (2009), pages 152–164, provides an excellent introductory discussion on difference and segment knotted spline variables.

To test the difference between S_1_2 and S_3_4, between S_3_4 and S_5_6, between S_5_6 and S_7_8, between S_7_8 and S_9_10, and between S_9_10 and S_11_12, we create five new variables by adding variables as follows:

SES-revised	D_1_2	D_3_4	D_5_6	D_7_8	D_9_10	D_11_12
1	1	1	0	0	0	0
2	2	2	0	0	0	0
3	3	3	1	0	0	0
4	4	4	2	0	0	0
5	5	2	1	0	0	0
6	6	3	2	0	0	0
7	7	4	3	1	0	0
8	8	5	4	2	0	0
9	9	6	5	3	1	0
10	10	7	6	4	2	0
11	11	8	7	5	3	1
12	12	9	8	6	4	2

$$D\_9\_10 = S\_9\_10 + S\_11\_12$$
$$D\_7\_8 = S\_7\_8 + D\_9\_10$$
$$D\_5\_6 = S\_5\_6 + D\_7\_8$$
$$D\_3\_4 = S\_3\_4 + D\_5\_6$$
$$D\_1\_2 = S\_1\_2 + D\_3\_4$$

The new variables are shown in the following table. Notice that D_1_2 is the same as the revised SES variable:

Going from right to left, the variable on the right is nested in the variable on the left. Nesting the variables in this way creates a second-order difference. The coefficient for the segment spline variable measures a first-order difference, how much more or less of the dependent variable results from a one-unit change in the independent variable. The new measure captures how much the coefficient for first segment spline variable is different from the coefficient for the second segment spline variable. I call this type of spline variable a "difference spline," and it is the type of spline variable that is commonly presented in discussions of spline variables in statistics texts. As an example of using segmented spline variables and difference spline variables, Table 7.4 shows the results from regressions that examine the effect of revised parental SES on math score.

Regression Modeling Using Spline Variables

Model 1 shows the effect of revised SES on math score. The coefficient shows that for every one unit of SES, math score increases by 1.78 units. In Model 2, I replace the revised SES variable with the segment spine variables. All but one of the segment splines is significant. Although the segmented splines do seem to increase as the level of SES increases, whether those increases are significant can be resolved with a direct test by using difference splines.

Table 7.4 Using Spline Variables to Model the Effect of the Revised SES Variable With Math Score as Dependent Variable

	Model		
	1	2	3
Independent Variable	B	B	B
SES-revised	1.78*	—	—
S_1_2	—	1.18	—
S_3_4	—	1.39*	—
S_5_6	—	1.31*	—
S_7_8	—	2.13*	—
S_9_10	—	2.24*	—
S_11_12	—	2.07*	—
D_1_2	—	—	1.18
D_3_4	—	—	.21
D_5_6	—	—	−.08
D_7_8	—	—	.82*
D_9_10	—	—	.11
S_11_12	—	—	−.17
Intercept	40.90	43.07	43.07
R^2	.158	.161	.161

*$p < .05$.

Model 3 includes the difference splines. To show the pattern of coefficients better, I include S_11_12 twice in the list of independent variables. Also, the D_1_2 variable is the same as the revised SES variable, but including D_1_2 shows better the logic of using difference splines.

The coefficient for D_1_2 in Model 3 is 1.18, and it is the same as the coefficient for S_1_2 in Model 2. The new information in Model 3 is the calculated differences. The coefficient for D_3_4 in Model 3 is .21, and it is the difference between the 1.39 and 1.18 coefficients in Model 2. The coefficient for D_5_6 in Model 3 is −.08, and it is the difference between the 1.31 and 1.39 coefficients in Model 2. The coefficient for D_7_8 in Model 3 is .82, and it is the difference between the 2.13 and 1.31 coefficients in Model 2. The coefficient for D_9_10 in Model 3 is .11, and it is the difference between the 2.24 and 2.13 coefficients in Model 2. Finally, the coefficient for the S_11_12 coefficient in Model 3 is −.17, and it is the difference between the 2.07 and 2.24 coefficients in Model 2.

Only the .82 coefficient in Model 3 is significant, which indicates that the 2.13 and 1.31 coefficients in Model 2 are significantly different from one another. This result indicates that the impact of each revised SES unit increases for unit values 7 and above, which raises the possibility that revised SES can be modeled with two variables rather than one.

I argue that dummy variable regression coefficients can be interpreted as first-order differences. The same goes for interval variables and segment spline variables. However, difference spline variables are second-order differences because they estimate differences between differences.

There is another way to estimate second-order differences in a spline model in addition to using difference spline variables. We can estimate second-order differences by applying the concept of nestedness in a different way.

The second way to consider differences between segment splines variables is to replace one segment spline with the original interval variable. In Table 7.5, Model 1 includes only the revised SES variable. Model 2 includes the segment splines. In Model 3, S_5_6 is replaced by the revised SES variable.

What happens in Model 3 of Table 7.5 is that the coefficients for the other segment splines are now the differences between the respective segment spline coefficients and the S_5_6 coefficient. This occurs because the other segment spline variables are nested in the revised SES variable. The revised SES coefficient in Model 3 is the same as the S_5_6 coefficient in Model 2 because the S_5_6 variable is the remainder when the other segment spline variables are nested in the revised SES variable.

In Model 3 of Table 7.5, the coefficients for S_1_2, S_3_4, and S_11_12 are not significant, which indicates that the coefficients for these variables in Model 2 are not significantly different from the coefficient for S_5_6 in Model 2. On the other hand, the coefficients for S_7_8 and S_9_10 in Model 3 are significantly different from zero. In Model 4, I replace S_7_8 with the original SES variable. The coefficients for S_3_4 and S_5_6 are significant, which indicates that the coefficients for these variables in Model 2 are significantly different from the coefficient for S_7_8 in Model 2. On the other hand, S-1_2, S_9_10, and S_11_12 in Model 4 are not significantly different from zero. These two sets of results can be used to formulate a simplified model.

| Table 7.5 | Using Spline Variables to Model the Effect of the Revised SES Variable With Math Score as Dependent Variable |

Independent Variable	Model					
	1	2	3	4	5	6
	B	*B*	*B*	*B*	*B*	*B*
SES-revised	1.78*	—	1.31*	2.13*	—	1.33*
S_1_2	—	1.18	−.13	−.96	—	—
S_3_4	—	1.39*	.08	−.74*	—	—
S_5_6	—	1.31*	—	−.82*	—	—
S_7_8	—	2.13*	.82*	—	—	—
S_9_10	—	2.24*	.93*	.11	—	—
S_11_12	—	2.07*	.76	−.06	—	—
S_1_6	—	—	—	—	1.33*	—
S_7_12	—	—	—	—	2.17*	.84*
Intercept	40.90	43.07	43.07	43.07	42.85	42.85
R^2	.158	.161	.161	.161	.161	.161

*$p < .05$.

The results from Models 3 and 4 in Table 7.5 indicate that some of the segment spline variables can be combined since the coefficients for the variables are not significantly different from one another. Combining the segment spline variables may produce a more parsimonious model that fits the data as well as the more complex model. Researchers value more parsimonious models over less parsimonious models because simplicity is preferred to complexity.

In Model 5 of Table 7.5, I combine S_1_2, S_3_4, and S_5_6 since the coefficients for the first two variables in Model 2 are not significantly different from the third. I also combine S_7_8, S_9_10, and S_11_12 because the coefficient in Model 2 for the first variable is not significantly different from the coefficients for the last two variables. The resulting coefficients for S_1_6 and S_7_12 in Model 5 are significant.

To test whether the coefficients for S_1_6 and S_7_12 in Model 5 of Table 7.5 are significantly different from one another and are not equal, I include the original SES variable and the S_7_12 variable in Model 6. Since S_7_12 is nested within revised SES, the coefficient for S_7_12 in Model 6 is the difference between the coefficients for

S_1_6 and S_7_12 in Model 5. The coefficient for S_7_12 in Model 6 is significant, indicating that the coefficients for S_1_6 and _7_12 in Model 5 are not equal.

The analysis in Table 7.5 indicates that the model with two spline variables might be the best model for capturing the effect of revised SES on math scores. However, we need to confirm this result statistically by using the F test.

First, I will check how Model 5 of Table 7.5 with two spline variables capturing SES compares with Model 2 which includes six spline variables. In this test, we will see whether the model with the two collapsed spline variables results in a model that fits as well as in a model with the six original spline variables. If Model 5 does not fit the data as well as Model 2, then Model 2 is the preferred model since it explains more of the variation in math scores.

The following is an F test for improvement in fit. The R^2 is .161108 for the small model and .161125 for the big model. The sample size is 20,133 for both models:

$$F = \frac{(.161125 - .161108)/4}{(1 - .161125)/(20,133 - 6 - 1)} = 0.10$$

The critical F statistic at the .05 level with $(4, \infty)$ degrees of freedom for the numerator and the denominator is 2.37. Since the calculated F statistic of 0.10 is less than the critical F statistic of 2.37, using two knotted spline variables did not lead to a significant change in R^2. This test indicates that Model 5 of Table 7.5 is preferred over Model 2. The smaller model fits the data as well as the larger model.

The last task in this exercise in spline modeling is to see whether Model 5 of Table 7.5 with two knotted spline variables capturing revised SES fits the data better than Model 1, which includes the original SES variable. If Model 5 does not fit the data significantly better than Model 1, then Model 1 is the preferred model since it is more parsimonious. I use an F test to test for improvement in fit (R^2 is .158208 for small model and .161108 for big model; sample size is 20,133):

$$F = \frac{(.161108 - .158208)/1}{(1 - .161108)/(20,133 - 2 - 1)} = 69.59$$

The critical F statistic at the .05 level with $(1, \infty)$ degrees of freedom for the numerator and denominator is 3.84. Since the calculated F statistic of 69.59 is greater than the critical F statistic of 3.84, using two knotted spline variables led to a significant improvement in R^2 over the model that used only the revised SES variable. In this example, the knotted spline variables were clearly not equal with one almost twice the size of the other. However, using a single variable to capture SES would not be completely misleading because the two knotted spline variables are both positive as is the coefficient for the single revised SES variable. If revised SES were just a control variable, using one variable for SES may not have much impact on the analysis. However, if the focus of the analysis was on the effect of SES, modeling SES with two spline variables would be preferred since all SES units do not have the same effect on the dependent variable.

Working With a Continuous Independent Variable

To create segment spline variables using a continuous variable as the base, the researcher must first decide on the number of segments to create. In the example that follows, I create segment splines for the continuous SES variable. I create six segments, each about equal length.

Each segment spline variable has an upper and a lower limit. First, I set all the segment spline variables equal to zero. Creating a particular segment spline variable involves subtracting the lower limit of the interval from a value if the value falls in the interval. If a value is larger than the upper limit of the interval, then the value of the segment spline variable is set equal to the size of the interval, which is the maximum value possible.

Notice that except for the first interval, the interval covers the part of the original variable greater than the lower limit and less than or equal to the upper limit. Also, the lower limit of the first segment is the minimum of all values and the upper limit of the last segment is the maximum. The following is SPSS syntax for creating the spline variables with S01, S02, S03, S04, S05, and S06 first set equal to zero:

```
IF -1.7501<=x2ses and x2ses<=-1.0835 S01=x2ses-(-1.7501).
IF x2SES>-1.0835 S01=.6666.
IF -1.0835<x2ses and x2ses<=-0.4167 S02=x2ses-(-1.0835).
IF x2SES>-0.4167 S02=.6668.
IF -0.4167<x2ses and x2ses<=0.2499 S03=x2ses-(-0.4167).
IF x2SES> 0.2499 S03=.6666.
IF 0.2499<x2ses and x2ses<=0.9167 S04=x2ses-( 0.2499).
IF x2SES>0.9167 S04=.6668.
IF 0.9167<x2ses and x2ses<= 1.5835 S05=x2ses-( 0.9167).
IF x2SES>1.5835 S05=.6668.
IF 1.5835<x2ses and x2ses<=2.2824 S06=x2ses-( 1.5835).
```

It is always good practice to check on the accuracy of variables after they are created. In this instance, the check is that the sum of segment spline variables plus the lower limit should equal exactly the value of the underlying continuous variable.

Table 7.6 shows the continuous segment spline variables with math scores as dependent. All six variables are significant in Model 1. The first three segment spline variables are smaller than the last three. Model 2 replaces S03 with SES, and Model 4 replaces S04 with SES. Based on the results of Models 3 and 4, Model 5 includes two combined variables, S_01_03 and S_04_06. Model 6 shows that 6.90 is significantly larger than 4.28.

Table 7.6 Using Spline Variables to Model the Effect of the SES Variable With Math Score as Dependent Variable

Independent Variable	Model					
	1	2	3	4	5	6
	B	*B*	*B*	*B*	*B*	*B*
SES	5.45*	—	3.96*	7.32*	—	4.28*
S01	—	2.90*	−1.06	−4.42*	—	—
S02	—	4.72*	.76	−2.60*	—	—
S03	—	3.96*	—	−3.36*	—	—
S04	—	7.32*	3.36*	—	—	—
S05	—	6.55*	2.59*	−.77	—	—
S06	—	5.60*	1.64	−1.72	—	—
S_01_03	—	—	—	—	4.28*	—
S_04_06	—	—	—	—	6.90*	2.62*
Intercept	51.14	43.85	50.78	56.66	43.13	50.62
R^2	.162	.165	.165	.165	.165	.165

*$p < .05$.

Example Using Logistic Regression

A researcher is interested in whether the effect of SES on chances of attending private high school is the same across all values of SES. The idea is that the effects of SES may diminish at the highest levels of SES. Table 7.7 shows the effect of revised SES in Model 1 and segment spline variables in Model 2. The coefficients for the segment splines appear to decrease at higher levels of SES. However, none of the difference splines in Model 3 are significant, which seems to indicate that the spline model is not necessary. However, the difference in −2 log-likelihood between Model 1 and Model 2 of 23.7 (16,115.1 − 16,091.4) is distributed chi-square and is larger than the critical chi-square with 5 degrees of freedom at the .05 level of significance of 11.07. Model 2 fits the data better than Model 1. When choosing between two models, the model with the significantly lower −2 log-likelihood is the preferred choice.

The researcher then estimates a simpler model than Model 2. Since the coefficients for S_9_10 and S_11_12 are less than the coefficient for S_7_8, the researcher creates two combined variables, S_1_8 and S_9_12. Model 4 shows that the coefficient for S_1_8 is larger than the coefficient for S_9_12 and that both are significantly different from zero. Model 5 shows that the difference between the two coefficients, −.15, is significantly different from zero.

Table 7.7 Using Spline Variables to Model the Effect of the Revised SES Variable with Private High School as Dependent Variable

Independent Variable	Model				
	1	2	3	4	5
	B	B	B	B	B
SES-revised	.39*	—	—	—	.44*
S_1_2	—	−.31	—	—	—
S_3_4	—	.60*	—	—	—
S_5_6	—	.44*	—	—	—
S_7_8	—	.41*	—	—	—
S_9_10	—	.32*	—	—	—
S_11_12	—	.22*	—	—	—
D_1_2	—	—	.31	—	—
D_3_4	—	—	.91	—	—
D_5_6	—	—	−.16	—	—
D_7_8	—	—	−.03	—	—
D_9_10	—	—	−.09	—	—
S_11_12	—	—	−.10	—	—
S_1_8	—	—	—	.44*	—
S_9_12	—	—	—	.29*	−.15*
Intercept	−4.22	−3.29	−3.29	−4.50	−4.50
−2 log-likelihood	16,115.1	16,091.4	16,091.4	16,096.3	16,096.3

*$p < .05$.

The difference in fit between Models 2 and 4 can be tested with a chi-square test. The difference in −2 log-likelihood between the two models of 4.9 (16,096.3 − 16,091.4) is distributed chi-square and is smaller than the critical chi-square with 4 degrees of freedom of 9.49. Model 4 fits the data as well as Model 2. When two models do not have significantly different −2 log-likelihood statistics, the model with fewer coefficients is the preferred choice. In this case, Model 4 is the preferred model. The analysis supports the idea of diminishing returns of SES on chances of attending private school. Higher units of SES have lower effects than lower units of SES.

Summary

The natural tendency when using an interval-level variable as an independent variable in a regression analysis is simply to enter the variable into the equation and estimate a linear effect. However, I suggest that researchers take time to consider the underlying nature of the linear effect. On one hand, if the coefficient for the interval variable is not significantly different from zero, then the lack of a significant effect may be due to a nonexistent relationship. However, it also could be due to constraining an effect of the independent variable on the dependent variable to be linear when it is actually not linear. On the other hand, the researcher might find a significant linear effect, but care must be taken in interpreting that effect. For example, the number of hours that a high school student studies for an examination may be positively related to test performance. An analysis using spline variables could show whether each additional hour has an equal effect or whether there are diminishing returns to additional hours once a certain number of hours studied is reached.

The standard approach to modeling possible nonlinearity in regression analysis is to add a squared term for the interval independent variable in question. This approach has some drawbacks such as the nonlinearity may not be best modeled by a squared term. Another drawback is that the coefficient for the squared term is not readily interpretable. In this book, I suggest that using spline variables is a good alternative to modeling linearity with squared terms because using spline variables is less restrictive in regard to the form of the nonlinearity and the results of analyzing nonlinearity using spline variables is directly interpretable.

There are two types of spline variables. Using segment spline variables allows for estimation of the effect of certain segments of the interval variable. Using difference spline variables allows for estimation of the size and significance of differences in the segment spline variables. As was the case for the standard interaction model and the within-groups interaction model, models with difference spline variables and segment spline variables work together to provide a complete picture of the linear effect being modeled.

Key Concepts

knotted spline variables: variables created by dividing an interval variable into segments such as for a variable ranging from 1 to 6 divided into segments of 1–2, 3–4, and 5–6; the term *knotted* is used because a graph of the coefficients for knotted spine variables will show that the segments connect to one another.

segment spline variables: spline variables that capture independent segments of an interval variable.

difference spine variables: spline variables that are nested one onto the other that allow for the testing of differences in the coefficients for segment spline variables.

***F* test:** a statistical test that allows for testing whether the difference in R^2 between two linear regression models is large enough to say that the overall fits of the two models are significantly different from one another.

chi-square test: a statistical test that allows for testing whether the difference in −2 log-likelihood between two logistic regression models is large enough to say that the overall fits of the two models are significantly different from one another.

Chapter Exercises

1. Replicate the spline variable analysis in Table 7.1 involving SES quartiles scores as the independent variable and math scores as the dependent variable. Use SESQ4_2, SESQ4_3, SESQ4_4, JSESQ4_2, JSESQ4_3, JSESQ4_4, and JSESQ4_2_SUM.

    The replication will first involve creating the dummy variables for SES quartile. The analysis will then involve calculating the "jump" variables. The final part of the replication will involve creating the table for the results of the regressions.

2. Replicate the spline variable analysis in Table 7.6 involving the continuous SES variable as the independent variable and math score as the dependent variable. Use X2SES, S01, S02, S03, S04, S05, S06, S_01_03, and S_04_06.

    The replication will first involve creating the segment variables as defined in the chapter. The final part of the replication will involve creating the table of results for the regressions.

3. Create four segment spline variables involving the continuous SES variable with the following four intervals: −1.7501 to −0.7501, −0.7501 to 0.2499, and 0.2499 to 1.2499, and 1.2499 to 2.2824. Each interval other than the last has a length of 1.0000. Use X2SES, and compute S1, S2, S3, S4, S12, and S34.

    The final part of the analysis using the newly created spline variables will involve creating a table of results that uses the linear regression procedure with math score as the dependent variable. In this analysis, you will need to figure out whether any of the segment spline variables can be combined.

Conclusion

Testing Research Hypotheses

Bivariate Hypothesis/No Controls

Although researchers often use regression analysis as a descriptive tool, another important use of regression analysis is to test research hypotheses. However, regression analysis can highlight only specific aspects of the relationships found in data. Thus, research hypotheses must take particular forms to be testable using regression analysis.

The simplest form of a research hypothesis is the bivariate hypothesis without controls. The bivariate hypothesis describes the form of the relationship between one independent variable and one dependent variable. The hypothesis takes different forms depending on whether the independent variable is an interval variable or a categorical variable and on whether the dependent variable is an interval variable or is a categorical variable.[1]

---

1. My approach to wording hypotheses is greatly influenced by Hage (1972) and, in particular, his Chapter 2 on Theoretical Statements.

Interval Independent/Interval Dependent

The higher the *X*, the higher (lower) the *Y*.

If the independent were socioeconomic status and the dependent variable were mathematics score, then the research hypothesis might be "the higher the parental socioeconomic status, the higher the student's mathematics score." If the dependent variable was the number of times a student skipped class, then the research hypothesis might be "the higher the parental socioeconomic status, the less often the student skips class"

Interval Independent/Dummy Dependent

The higher the *X*, the more (less) likely to be *Y*.

With a two-category, dummy-dependent variable, the dependent variable is not measuring the amount of something but is measuring the relative chances of being in one category as opposed to another category. If the independent were socioeconomic status and the dependent variable were attended private school, then the research hypothesis might be "the higher the parental socioeconomic status, the more likely the student attends private school."

Dummy Independent/Interval Dependent

Those in Category A will be higher (lower) on *Y* than those in Category B.

The nature of the hypotheses involving a dummy independent variable and an interval dependent variable depends on how many categories are in the independent variable. If there are two categories for the independent variable, then there will be one hypothesis. If there are more than two categories for the independent variable, then there may be multiple hypotheses. If the independent variable were a dummy variable for female and the interval dependent variable were math scores, then the hypothesis might be "females have lower math scores than males." Stating the hypothesis in this way emphasizes female disadvantage. However, the researcher has a choice as to which category of the dummy variable to emphasize. An alternative hypothesis might be "males have higher math scores than females." This hypothesis would emphasize male advantage.

A researcher will use more than one dummy variable to represent an independent variable when the variable has more than one category. For example, the family structure of the parental household might be divided into two biological parents, biological parent/ stepparent, single parent, and other family. A key decision is which category is the contrast group. This is the category that all the other categories are compared against. Often in research using family structure as the independent variable, the two-biological-parent category is used as the contrast group. However, if the research emphasized the disadvantage of those in single-parent families, then perhaps the single-parent category might be used as the contrast group.

If the two-biological-parent category is the contrast group, then there may be two hypotheses. One hypothesis might be "those in biological-parent/stepparent families have lower math scores than those in two-biological-parent families." The other hypothesis might be "those in single-parent families have lower math scores than those in two-biological-parent families."

In the case of the family structure variable, the researcher will likely not suggest a hypothesis for the "other family" category. This category is constructed so that the other three categories are more uniform. The other three categories would not be as uniform if the "other family" category was combined with any one of them. However, the "other family" category may be so diverse that analyzing the difference between it and the two-biological-parent category would not be meaningful. The coefficient for the dummy variable representing this category would be estimated but not analyzed. However, including this variable is important for the clarity of the analysis.

Hypotheses involving categories of a categorical variable do not necessarily have to work in the same direction. For race/ethnicity, a researcher might construct the variable to have the following five categories: non-Hispanic White, Hispanic, African American, Asian, and other race/ethnicity. By using non-Hispanic White as the contrast group, the research could generate the following three hypotheses: "Hispanics have lower math scores than non-Hispanic Whites," "African Americans have lower math scores than non-Hispanic Whites," and "Asians have higher math scores than non-Hispanic Whites." The difference for the other race/ethnicity category would not be analyzed.

If the African American category were the contrast group, then the hypotheses might be that there is no difference between that group and Hispanics but that the group is lower than non-Hispanic Whites and Asians. The coefficient for the difference between the African American category and the other category would not be analyzed.

Dummy Independent/Dummy Dependent

Those in Category A are more likely to be Y than those in Category B.

As mentioned, concerning the situation when there is a two-category, dummy-dependent variable, the dependent variable is not measuring the amount of something but is measuring the relative chance of being in one category as opposed to another category. If the independent were a dummy variable for college-graduate parent and the dependent variable were attended private school, then the research hypothesis might be "those respondents who have a college graduate parent are more likely to attend private school than those respondents who do not have a college graduate parent."

Using an independent variable with more than two categories requires a decision about the contrast group. Although the contrast group is sometimes called the "excluded category," the choice of contrast group is the key to the analysis as it defines the differences being analyzed.

Bivariate Hypothesis/Unanalyzed Controls

The reason that regression analysis is used in survey research to such a great extent is that in survey research, the "experimental treatment" is not randomly assigned like in an experiment. In the High School Longitudinal Study data used in this book, respondents are not randomly assigned to situations such as attending private school or not. Students in private school and those students not in private school differ on other characteristics related to math scores rather than private school attendance, and researchers use regression analysis to control for these factors.

There are two key types of control variables. One type is a correlated factor, and the other type is an underlying mechanism. In the case of private school, those who have parents with more education and higher family income are more likely to attend than those without. Parental education and family income have positive influences on math scores for students whether in private school or not. Thus, the researcher needs to control for these correlated factors to get a more "pure" effect of private school on math scores.[2]

Underlying mechanism control variables are those that clarify the effect of the independent variable of interest on the dependent variable. Often, the independent variable of interest is a more abstract variable that can be viewed as capturing a "bundle" of effects. The analytical goal of controlling is to understand this overall effect.

The bivariate hypothesis with unanalyzed controls will usually look exactly like the bivariate hypothesis with no controls. The difference is the researcher will state that the hypothesis expresses the expected relationship once correlated factors are taken into account.

For example, researchers may be interested in whether students in private high school score higher on college admission tests than student in public high schools. The researcher knows from previous research that students in private high schools are more likely to have parents with bachelor and graduate degrees and have parents with higher incomes than students in public high schools. The researcher wants to estimate the "pure" effect of private school controlling for other variables.

In this situation, the researcher might specify the following hypothesis: "Those students attending private high schools score higher on college admission tests than those students who attend public high schools." He or she might not mention control variables until the methods section of the paper and then mention at that point that parental education and family income are controlled in the analysis. On the other hand, the researcher may discuss in the hypotheses section of the paper the other variables that affect private school and, thus, must be controlled.

When a researcher reports the results for a bivariate hypothesis with unanalyzed controls, the researcher may just show one big model with the independent variable of interest and the controls. Sometimes the researcher may show a small model with the independent variable of interest and the big model, which also includes the control

---

2. Stinchcombe (1968), pages 60–79, discusses demographic explanations of social phenomenon and provides a thorough examination of the logic of control that is used in the present book.

| Table 8.1 | Linear Regression of Independent Variables on Math Score |

Independent Variable	B
Private	1.78*
Family Income	.13*
2-Yr. Degree	1.14*
4-Yr. Degree	4.79*
Grad. Degree	7.34*
Bio./Step.	−1.33*
Single	−.82*
Other Fam.	−1.87*
Intercept	47.92
R^2	.157

*$p < .05$.

variables. The model in Table 8.1 is one big model and shows the effect of private school on math scores holding constant family income, parental education, and family structure.

Hypotheses with unanalyzed controls are appropriate in situations where the researcher is examining the effect of a variable that has not been examined in previous research. Quasi-experimental designs often involve hypotheses of this type. In this type of design, the situations that receive an intervention into a social process are compared with the situations that do not receive an intervention.

For example, a researcher may develop a college awareness program for low-income, high-school juniors. The researcher may then observe whether high-school juniors in schools who receive the treatment are more likely to attend a college immediately after high-school graduation than are high-school juniors who are in schools where the program is not available. Since the students in the schools where the program was carried out may not have the same background characteristics as the student in the schools where the program was not carried out, these background factors need to be controlled.

Another situation where a hypothesis with unanalyzed controls may be used would be where a variable with a well-known effect is examined by using an unexplored dependent variable. For example, although the effects of socioeconomic status variables such as parental education and family income have been thoroughly studied in regard to educational attainment, less is known about the effects of these variables on what happens when a student attends college. The researcher may be interested in whether students study abroad as part of their undergraduate experience. The researcher might examine the effects of parental socioeconomic status on the propensity to study abroad holding constant whether the university is public or private.

Bivariate Hypothesis/Analyzed Controls

There are two situations where controls are analyzed in explaining the effect of an independent variable of interest. In the first situation, the researcher considers correlated factors to determine the degree to which the effect of the independent variable of interest is a result of the correlated factors. In the second situation, the researcher considers variables that capture underlying mechanisms to determine the degree to which the effect of the independent variable of interest is a result of underlying mechanisms.

The researcher may use one of the control modeling approaches discussed earlier to consider the roles that the control variables play in explaining the effect of the independent variable of interest. The control variables may be a mixture of correlated factors and underlying mechanism control variables.

The wording of a hypothesis with analyzed controls involves stating how the relationship between X and Y changes when Z is controlled. The hypothesis requires defining how X affects Y, how X affects Z, and how Z affects Y. This hypothesis can be adjusted to accommodate dummy variables.

Hypothesis for Analyzed Controls

Part of the reason that the higher the X, the higher (lower) the Y is that the higher the X, the higher (lower) the Z and the higher the Z, the higher (lower) the Y.

For example, Table 8.2 shows how controlling for family income, parental education, and family structure explains part of the effect of attending a private school on math scores. The hypothesis about controlling for parental education would be as follows:

> Part of the reason that those who attend private school have higher math scores than those who do not attend private school is that those who attend private school are more likely to have parents with higher education than those who do not attend private school and those whose parents have more education score higher in math than those whose parents have less education.

Hypothesis Involving Interactions

A common situation in which interactions are used in an analysis is the situation when at the end of the discussion of an analysis, there will be a segment on the analysis of interactions. This discussion comes across as an afterthought because in the section of the paper that deals with research questions, often there is no mention of interaction effects.

Table 8.2 One-at-a-Time Without Controls With Math Score as Dependent Variable

Independent Variable	Model				
	1	2	3	4	5
	B	*B*	*B*	*B*	*B*
Private	5.06*	3.14*	2.66*	4.61*	1.78*
Family Income	—	.23*	—	—	.13*
2-Yr. Degree	—	—	1.32*	—	1.14*
4-Yr. Degree	—	—	5.57*	—	4.79*
Grad. Degree	—	—	8.72*	—	7.34*
Bio./Step.	—	—	—	−2.04*	−1.33*
Single	—	—	—	−2.30*	−.82*
Other Fam.	—	—	—	−3.25*	−1.87*
Intercept	50.80	49.01	47.88	51.91	47.92
R^2	.035	.084	.140	.049	.157

*$p < .05$.

The reason that a researcher might add an interaction analysis at the end of an analysis may be a reviewer or thesis committee member who suggests that the researcher might check for interactions or just the researcher wanting to be thorough and to examine all possibilities in regard to modeling the effects of a variable.

I suggest strongly that the researcher should not include interactions in an analysis unless the researcher has addressed the issue of interaction when discussing research questions. The problem with adding an interaction analysis at the end of an analysis is that the researcher has not justified the analysis with an examination of previous research and with a careful discussion of how the interactions are expected to work.

Most researchers would not present an analysis of the effects of an additive variable on dependent variables without some theoretical rationale. The same should hold true for interactions. There should also be a theoretical rationale.

For example, a researcher examines the effect of attending a private high school versus a public high school on math scores. At the end of the analysis, the researcher examines the interaction between gender and private high school attendance on math scores without previous discussion in the research questions section of the paper. The problem with examining interactions in this manner is that the meaning of the interaction of gender with private high school attendance is not about whether there is an effect of private high school on math scores, but it is about whether that effect differs between males and females.

Understanding the effect of attending a private high school on math scores involves understanding what happens in a private high school as opposed to a public high school that would encourage higher math scores. These differences might include smaller class sizes, more skillful math teachers, and higher aptitude students in private high schools as compared with public high schools. The interaction between gender and private high school is not about whether smaller class sizes, more skillful math teachers, or higher aptitude students have an impact on math scores, but it is about whether females benefit from these factors as much as males do. For example, perhaps math teachers who are more skillful are more sensitive to gender stereotypes, which then leads to female students more effectively using their innate math abilities.

The argument about how the effect of a variable differs between two or more groups and is very different than the argument about how one variable affects another variable. The interaction argument is about how a social process differs between two or more groups. The additive argument is about how that social process works for everyone.

The wording of a hypothesis for interactions involves stating how the relationship between X and Y differs between groups. The hypothesis requires defining how X affects Y and how that relationship differs between Groups A and B. This hypothesis can be adjusted to accommodate dummy variables used to model the independent variables or the dependent variable.

Hypothesis for Interactions

The effect of X on Y is larger (smaller) for Group A than for Group B.

For example, Table 8.3 investigates the interaction between family structure and parental education. The idea is that those respondents in biological-parent/stepparent and single-parent families might receive less benefit from having a college-graduate parent than those who live with two-biological-parent families. This might occur as a result of less parental attention in stepparent and single-parent families than in two-biological-parent families.

The hypothesis about the interaction between family structure and parental education would be as follows:

The effect of having a college-graduate parent on math scores is higher for those living with two biological parents than for those living with a biological parent/stepparent or with a single parent.

Model 3 in Table 8.3 shows that the effect of having a college-graduate parent on math scores is higher for those living with two biological parents than for those living with a biological parent/stepparent or with a single parent. Model 2 shows that the differences in these effects are significant.

Table 8.3 Interactions Between Family Structure and College-Graduate Parent With Math Score as Dependent Variable

	Model		
	1	2	3
Independent Variable	B	B	B
Bio./Step.	−1.70*	−1.09*	−1.09*
Single	−1.39*	−1.08*	−1.08*
Other Fam.	−2.43*	−2.32*	−2.32*
College-Graduate Parent	6.65*	7.01*	—
Two Bio. × College Grad.	—	—	7.01*
Bio./Step. × College Grad.	—	−1.47*	5.54*
Single × College Grad.	—	−.76*	6.25*
Other Fam. × College Grad.	—	−.13	6.88*
Intercept	49.45	49.27	49.27
R^2	.124	.125	.125

*$p < .05$.

Hypothesis Involving Nonlinearity

Issues of nonlinearity in research may occur when a control variable may have a non-linear effect on the dependent variable. In this situation, the researcher may mention in the methods section of the paper that the variable will be included in the analysis and the way that the variables will be modeled. A common example of this situation is when age is included in a regression model as a control variable. The relationship between age and certain dependent variables is nonlinear and takes a U-shape or an inverted U-shape. For example, for those older than 18 years old, earned income rises through the 20s and then falls starting in the 50s. This occurs because as more individuals in an age group finish their educations and find jobs, the income for age groups goes up. Eventually, as individuals begin retiring from paid employment, the earned income by age goes down. This nonlinearity can be modeled by including an interval variable for age along with that variable squared, by modeling age using a set of dummy variables, or by using spline variables.

The drawback to modeling nonlinearity with a squared term is that the nature of the nonlinearity is not immediately obvious when examining the coefficients. Alternatively, the nonlinearity could be modeling in a way where the resulting coefficients are more readily interpretable than the coefficient for a squared term. Using a series of dummy variables to capture age or using spline variables results in coefficients that are immediately interpretable.

In regard to the independent variable of interest in an analysis, there are two situations where possible nonlinearity might be considered. The first is as a check on linearity assumptions involving the independent variable of interest. In this case, the researcher might only mention in the methods section of the paper that possible nonlinearity was considered but not found.

It is particularly important to consider nonlinearity when the finding about the effect of the independent variable of interest has implications for social policy. The reason it is important to consider nonlinearity in this case is that the linearity assumption suggests that each additional unit of the independent variable produces the same effect on the dependent variable. If this is not the case, any intervention program created to deal with the issue will need to be adjusted to accommodate the nonlinearity.

For example, suppose a researcher finds that the number of hours that a parent reads to a 5-year-old child has a positive effect on the child's reading ability. The researcher decides to check for possible threshold effects and diminishing returns using spline variables. A threshold effect would be that there is a minimum number of hours being read to be necessary for later reading to be affected. Diminishing returns means that after a certain number of hours, being read to more hours has little additional effect.

The second situation where possible nonlinearity might be considered is where the research question specifically addresses nonlinearity. Although hypotheses addressing nonlinearity are not common in social science research, there is much potential for new insights in exploring them.

The wording of a hypothesis for nonlinearity can take on several different forms since there is a variety of possible types of nonlinearity. A hypothesis dealing with nonlinearity involves stating how the relationship between X and Y differs across the range of X. To consider nonlinearity using splines, the independent variable must be interval. These hypotheses can be adjusted to accommodate dummy variables used as the dependent variable. I discuss three possible hypotheses involving nonlinearity.

Hypothesis for Nonlinearity: Threshold Effect

X is not related to Y until X is larger than value W.
When X is greater than W, then the higher the X, the higher the Y.

A threshold effect occurs in a situation in which the independent variable has to reach a certain level before it affects the dependent variable. For example, family income might have to reach a certain level before it has an effect on chances of attending private high school.

Diminishing returns occurs in a situation in which once the independent variable reaches a certain level, then additional units have less effect on the dependent variable. For example, once family income reaches a certain level, then higher levels of family income have less effect on chances of attending private high school.

Hypothesis for Nonlinearity:
Diminishing Returns

Up to when X reaches a value W, the higher the X, the higher the Y.
After W, Y does not continue to increase as much with additional units of X.

An inverted U-shaped relationship is when as X increases, Y increases up to a certain point, and then after that point, as X increases, then Y decreases. For example, earnings for adults between the ages of 18 and 70 increase from age 18 to the 50s and then decrease from the 50s to age 70. As individuals finish schooling, earnings increase with age, and in later years, individuals begin retiring from work and earnings decrease.

Hypothesis for Nonlinearity:
Inverted U-Shaped Relationship

Up to when X reaches a value W, the higher the X, the
higher the Y. After W, the higher the X, the lower the Y

There is a strong relationship between parental socioeconomic status (SES) and a student's math scores. Those students whose parents have more education, income, and occupational status score higher in math than those who have less of those factors. Table 8.4 shows an analysis where the researcher is considering whether there is a threshold effect for SES. That is, SES may have lower effects on math scores until SES reaches a certain level, and then the effect will be higher. The analysis in Table 8.4 did find a threshold effect as the effect of SES was higher at higher levels of SES than at lower levels.

Final Comments

The challenge in using regression analysis for social science research is to posit a model that addresses well-reasoned research questions. The objective of this book is to provide techniques for using regression analysis in this manner. Regression modeling is more than a way to analyze data, it is a way to answer research questions.

Regression analysis involves estimating regression equations that examine the relationship between independent and dependent variables. In this book, I have defined regression modeling as the in-depth examination of a regression coefficient that uses control models, interactions, or spline variables. Although a single regression equation allows for a test of a simple hypothesis involving an independent and a dependent variable, regression modeling allows for the examination of more complex hypotheses involving correlated factors, intermediate mechanisms, conditional effects, and nonlinearity.

	Table 8.4 Using Spline Variables to Model the Effect of the Revised SES Variable With Math Score as Dependent Variable			

	Model			
	1	**2**	**3**	**4**
Independent Variable	***B***	***B***	***B***	***B***
SES-revised	1.78*	—	—	1.33*
S_1_2	—	1.18	—	—
S_3_4	—	1.39*	—	—
S_5_6	—	1.31*	—	—
S_7_8	—	2.13*	—	—
S_9_10	—	2.24*	—	—
S_11_12	—	2.07*	—	—
S_1_6	—	—	1.33*	—
S_7_12	—	—	2.17*	.84*
Intercept	40.90	43.07	42.85	42.85
R^2	.158	.161	.161	.161

*$p < .05$.

This book uses what I call a "discrete" approach, In the discrete approach, the conceptual basis is not a cloud of points illustrating the relationship between a continuous X variable and a continuous Y variable but instead focuses on discrete groups and how those groups differ on a dependent variable. The discrete approach is useful for social science researchers because it allows the idea of social groups to be at the forefront of research that uses regression analysis and modeling.

Summary

Although this chapter appears at the end of this book on regression modeling, the creation of research questions should come first in the process of regression modeling. Sometimes these research questions will be formally written, but at other times, they will be just thoughts in the mind of the researcher. Either way, the basis of a regression model or set of regression models is a researcher's question. That research question is best thought of in terms of a testable hypothesis.

There are four main situations where we test hypotheses by using regression models. First, we are interested in whether a particular independent variable has an effect on a dependent variable and the nature of that effect. In this case, we are interested in the significance and size of the effects of the independent variable. Second, we are interested in the degree to which control variables explain the effect of an independent variable of interest. This analysis involves control modeling. Third, we are interested in whether the effect of an independent variable is conditional on the level of a second independent variable. This analysis will involve interaction modeling. Fourth, we are interested in the degree to which the effect of an interval independent variable on a dependent variable is linear. This analysis will involve the use of spline variables.

Although sitting at a desk and using a computer to estimate regression models may not seem as exciting as going out into the field and observing society at work or talking to people about their lives, the thrill of discovery is always near. Creating research questions and using regression analysis to find answers can be an exciting endeavor as the researcher seeks to advance knowledge, whether in a small way or in a big way. I wish you well in your search for answers to questions.

Key Concepts

bivariate hypothesis/no controls: hypothesis that focuses on the effect of a particular independent variable on a dependent variable.

bivariate hypothesis/unanalyzed controls: hypothesis that focuses on the effect of a particular independent variable on a dependent variable with other independent variables controlled but with no analysis of the impact of the control variables on the effect of the independent variable of interest.

bivariate hypothesis/analyzed controls: hypothesis that focuses on the effect of a particular independent variable on a dependent variable with other independent variables controlled and with analysis of the impact of the control variables on the effect of the independent variable of interest by using a control modeling approach.

hypothesis involving interactions: hypothesis that focuses on the degree to which the effect of a particular independent variable on a dependent variable is dependent on the level of a second independent variable.

hypothesis involving nonlinearity: hypothesis that focuses on the degree to which the effect of a particular independent variable on a dependent variable follows a linear pattern.

Chapter Exercises

1. Family income and math scores are both interval variables. Researchers have shown that those with higher family incomes score higher on math than those with lower family incomes. State a hypothesis for this relationship.

2. Family income is an interval variable, and private high school is a dummy variable. Researchers have shown that those with higher family incomes are more likely to attend private high school than those with lower family incomes. State a hypothesis for this relationship.

3. Single-parent family and private high school are both dummy variables. Researchers have shown that those in single-parent families are less likely to attend private high school than those not in single-parent families. State a hypothesis for this relationship.

4. Researchers have shown that those attending private high schools have higher family incomes than those not attending private schools and that those who attend private high school have higher math scores than those who do not. State a hypothesis with attending private high school as the independent variable of interest and family income as the control variable.

5. Family income and math score are interval variables, and private high school is a dummy variable. Researchers have shown that the effect of family income on math scores is greater for those in private high school than for those not. State a hypothesis about the interaction of attending private high school with family income.

Appendix A

Creating the HSLS Data File

Accessing and Downloading the HSLS data

The data for the analysis presented in this book are from the High School Longitudinal Study of 2009 (HSLS). A public use version of this data is available at the following website supported by the National Center for Education Statistics:

https://nces.ed.gov/edat/

This site provides a connection to the Education Data Analysis Tool (EDAT) which allows access to HSLS data. The dataset used in this book includes the following variables from the public use HSLS dataset:

STU_ID	student ID
W2STUDENT	first follow-up analytical weight
X2SEX	student's sex
X2RACE	student's race/ethnicity composite
X2TXMTSCOR	mathematics standardized score
X2PAREDU	parents' highest level of education
X2PARPATTERN	relationship pattern
X2FAMINCOME	total family income 2011
X2SES	socioeconomic status composite
X2SESQ5	quintile coding of X2SES
X2CONTROL	school control

The first wave of the study occurred in the fall of 2009 when the students were freshman in high school, and the second wave occurred in the spring of 2012 when the students were juniors in high school.

Creating the HSLS Extract

The first step in creating an extract from the larger HSLS dataset was to download the dataset using the EDAT tool provided by the National Center for Education Statistics.

Once the larger dataset was downloaded, the next step was to create an extract for analytical purposes. The analysis in this book uses data for students gathered in the second wave of the study in 2012. W2STUDENT is the student weight used to analyze data from all first follow-up responding students. To select the students who participated in that wave, I chose only those cases where W2STUDENT > 0. There were 23,415 students in the full dataset, and after selecting on W2STUDENT, there were 20,594.

The next step was to exclude those students who had a value of 1 or 2 on X2CONTROL (school control). After selecting only students who had values of W2STUDENT > 0, some students were categorized as "missing" (value = −6) on X2CONTROL, and some were categorized as "component not applicable" (value = −9).

After deleting those students who were −6 or −9 on X2CONTROL, the final dataset included 20,133 students. There are not any missing or not applicable values on any of the variables in the final dataset. The absence of missing or not applicable values is not typical of datasets that researchers actually use. However, the focus of this book is on regression modeling, and a "clean" dataset was constructed to facilitate a smooth discussion of regression modeling.

Access helpful study tools and resources—all in one place!

The data and statistical program code for this book are also available on the SAGE website at study.sagepub.com/wojtkiewicz.

Appendix B

Codebook for HSLS Variables

The following shows information for the variables taken directly from the EDAT codebook for the HSLS data:

STU_ID

Student ID

Student identifier assigned for all base-year eligible students (including respondents, nonrespondents, and questionnaire-incapable students). IDs randomly assigned from 10001 to 35206 across all students.

W2STUDENT

W2 First follow-up student analytic weight

Student weight used in analysis of data from all first follow-up responding students.

X2SEX

X2 Student's sex

X2SEX is the sample member's sex.

1   Male

2   Female

X2RACE

X2 Student's race/ethnicity-composite

X2RACE characterizes the sample member's race/ethnicity by summarizing the following six dichotomous race/ethnicity composites: X2HISPANIC, X2WHITE, X2BLACK, X2ASIAN, X2PACISLE, and X2AMINDIAN.

1   Amer. Indian/Alaska Native, non-Hispanic

2   Asian, non-Hispanic

3   Black/African-American, non-Hispanic

4   Hispanic, no race specified

5   Hispanic, race specified

6   More than one race, non-Hispanic

7   Native Hawaiian/Pacific Islander, non-Hispanic

8   White, non-Hispanic

X2TXMTSCOR

X2 Mathematics standardized theta score

The math standardized T score provides a norm-referenced measurement of achievement, that is, an estimate of achievement relative to the population (fall 2009 ninth graders) as a whole. It provides information on status compared to peers (as distinguished from the item-response theory-estimated percent-correct score, which represents status with respect to achievement on a particular criterion set of test items). The standardized T score is a transformation of the IRT theta (ability) estimate, rescaled to a mean of 50 and standard deviation of 10.

Continuous

X2PAREDU

X2 Parents'/guardians' highest level of education

Indicates the highest level of education achieved by either parent 1 (P1) or parent 2 (P2).

1   Less than high school

2   High-school (HS) diploma or GED or alternative HS credential

3   Certificate/diploma from occupational training

4   Associate's degree

5   Bachelor's degree

6   Master's degree

7   PhD/MD/Law/other high-level professional degree

X2PARPATTERN

X2 P1-P2 relationship pattern

This variable indicates (1) whether there are one or two parents in the sample member's home, (2) the relationship of those parent(s) to the sample member, and (3) if there are two parents in the home, the relationship of those parents to each other.

1   Two bio/adoptive parents

2   Bio/adoptive mother and non-bio/adoptive partner

3   Bio/adoptive mother and non-partner guardian

4   Bio/adoptive father and non-bio/adoptive partner

5   Bio/adoptive father and non-partner guardian

6   Two other guardians

7   Bio/adoptive mother only

8   Bio/adoptive father

9   Other female guardian only

10   Other male guardian only

11   Student lives with P1/P2 less than half the time

X2FAMINCOME

X2 Total family income from all sources 2011

X2FAMINCOME is a categorical variable that indicates the sample member's family income from all sources in 2011, as reported by the parent questionnaire respondent.

1   Family income less than or equal to $15,000

2   Family income > $15,000 and ≤ $35,000

3   Family income > $35,000 and ≤ $55,000

4   Family income > $55,000 and ≤ $75,000

5   Family income > $75,000 and ≤ $95,000

6   Family income > $95,000 and ≤ $115,000

7   Family income > $115,000 and ≤ $135,000

8   Family income > $135,000 and ≤ $155,000

9   Family income > $155,000 and ≤ $175,000

10   Family income > $175,000 and ≤ $195,000

11   Family income > $195,000 and ≤ $215,000

12   Family income > $215,000 and ≤ $235,000

13   Family income > $235,000

X2SES

X2 Socioeconomic status composite

Description:
This composite variable is used to measure a construct for socioeconomic status.

Continuous

X2SESQ5

X2 Quintile coding of X2SES composite

This variable is the quintile of X2SES, weighted using the student weight (W2STUDENT).

1   First quintile (lowest)

2   Second quintile

3   Third quintile

4   Fourth quintile

5   Fifth quintile (highest)

X2CONTROL

X2 School control

X2CONTROL identifies the sample member's first follow-up school as being a Public, Catholic, or Other Private School, as indicated in the source data for sampling: the Common Core of Data (CCD) 2011–2012 and the Private School Survey (PSS) 2011–2012.

1   Public

2   Catholic or other private

Appendix C

Creating HSLS Variables

Researchers use a variety of statistical packages to analyze data including widely available statistical packages such as IBM® SPSS® Statistics* (SPSS Corporation, Chicago, IL) and SAS (SAS Institute, Inc., Cary, NC) and more specialized packages such as STATA (StataCorp, College Station, TX) and R (R Development Core Team).

To preserve space, I will explain how to recode the variables that I referenced in this book by using SPSS syntax. I believe that the most important information concerning recoding variables is the logic behind the recoding. Once that logic is understood, creating recodes in other statistical software packages is a straightforward task.

In the following, I provide for each variable a short description, the SPSS recode syntax, and an example method for checking the recoding.

Chapter 2. Basic Statistical Procedures

PRIVATE
(two categories: private, public)

RECODE x2control (2=1)(1=0) INTO private.

x2control * private Crosstabulation

Count

		private		Total
		.00	1.00	
x2control	1	16797	0	16797
	2	0	3336	3336
Total		16797	3336	20133

*SPSS is a registered trademark of International Business Machines Corporation.

TOP 25% MATH
(two categories: top 25%, not top 25%)

RECODE x2txmtscor (Lowest thru 57.9900=0)
(57.9901 thru Highest=1) INTO highmath.

highmath

		Frequency	Percent	Valid Percent	Cumulative Percent
	.00	15108	75.0	75.0	75.0
Valid	1.00	5025	25.0	25.0	100.0
	Total	20133	100.0	100.0	

FAMILY STRUCTURE
(four categories: two biological parents, biological parent/stepparent, single parent, other)

RECODE x2parpattern (1=1) (2=2) (3=4) (4=2) (5 thru 6=4)
(7 thru 8=3) (9 thru 11=4) INTO famstruct.

x2parpattern * famstruct Crosstabulation

Count

		famstruct				Total
		1.00	2.00	3.00	4.00	
	1	11443	0	0	0	11443
	2	0	2275	0	0	2275
	3	0	0	0	487	487
	4	0	537	0	0	537
	5	0	0	0	127	127
x2parpattern	6	0	0	0	355	355
	7	0	0	3528	0	3528
	8	0	0	586	0	586
	9	0	0	0	169	169
	10	0	0	0	34	34
	11	0	0	0	592	592
Total		11443	2812	4114	1764	20133

Chapter 3. Regression Modeling Basics

NOT PRIVATE
(two categories: public, private)

RECODE x2control (1=1)(2=0) INTO nprivate.

x2control * nprivate Crosstabulation

Count

		nprivate		Total
		.00	1.00	
x2control	1	0	16797	16797
	2	3336	0	3336
Total		3336	16797	20133

FAMILY STRUCTURE DUMMIES
(one dummy variable for each category for each of four categories: two biological parents, biological parent/stepparent, single parent, other family)

RECODE x2parpattern (1=1) (else=0) INTO twopar.
RECODE x2parpattern (2=1) (4=1) (else=0) INTO step.
RECODE x2parpattern (7 thru 8=1) (else=0) INTO single.
RECODE x2parpattern (3=1) (5 thru 6=1) (9 thru 11=1) (else=0) INTO famoth.

x2parpattern * famoth Crosstabulation

Count

		famoth		Total
		.00	1.00	
x2parpattern	1	11443	0	11443
	2	2275	0	2275
	3	0	487	487
	4	537	0	537
	5	0	127	127
	6	0	355	355
	7	3528	0	3528
	8	586	0	586
	9	0	169	169
	10	0	34	34
	11	0	592	592
Total		18369	1764	20133

(Note: there are similar tables possible for twopar, step, and single)

SES QUINTILE DUMMIES
(one dummy variable for each category of SES quintile)

RECODE x2sesq5 (1=1) (else=0) INTO sesq51.
RECODE x2sesq5 (2=1) (else=0)INTO sesq52.
RECODE x2sesq5 (3=1) (else=0)INTO sesq53.
RECODE x2sesq5 (4=1) (else=0)INTO sesq54.
RECODE x2sesq5 (5=1) (else=0)INTO sesq55.

x2sesq5 * sesq52 Crosstabulation
Count

		sesq2		Total
		.00	1.00	
x2sesq5	1	3056	0	3056
	2	0	3566	3566
	3	3798	0	3798
	4	4426	0	4426
	5	5287	0	5287
Total		16567	3566	20133

(Note: there are similar tables possible for sesq1, sesq3, sesq4, and sesq5)

PARENTAL EDUCATION DUMMIES (SIX CATEGORIES)
(one dummy variable for each of six categories: <high school, high school only, 2 year degree, 4 year degree, MA/MS, PhD/MD)

RECODE x2paredu (1=1) (else=0) INTO pared1.
RECODE x2paredu (2=1) (else=0) INTO pared2.
RECODE x2paredu (3 thru 4=1) (else=0) INTO pared3.
RECODE x2paredu (5=1) (else=0) INTO pared4.
RECODE x2paredu (6=1) (else=0) INTO pared5.
RECODE x2paredu (7=1) (else=0) INTO pared6.

x2paredu * pared2 Crosstabulation
Count

		pared2		Total
		.00	1.00	
x2paredu	1	1028	0	1028
	2	0	6112	6112
	3	960	0	960
	4	3211	0	3211
	5	5002	0	5002
	6	2549	0	2549
	7	1271	0	1271
Total		14021	6112	20133

(Note: there are similar tables possible for pared1, pared3, pared4, pared5, and pared6)

PARENTAL EDUCATION (SIX CATEGORIES)

(six categories: <high school, high school only, 2 year degree, 4 year degree, MA/MS, PhD/MD)

RECODE x2paredu (1=1) (2=2) (3 thru 4=3) (5=4) (6=5) (7=6) INTO paredsix.

x2paredu * paredsix Crosstabulation

Count

		paredsix						Total
		1.00	2.00	3.00	4.00	5.00	6.00	
x2paredu	1	1028	0	0	0	0	0	1028
	2	0	6112	0	0	0	0	6112
	3	0	0	960	0	0	0	960
	4	0	0	3211	0	0	0	3211
	5	0	0	0	5002	0	0	5002
	6	0	0	0	0	2549	0	2549
	7	0	0	0	0	0	1271	1271
Total		1028	6112	4171	5002	2549	1271	20133

SES DECILE
(values: first decile through tenth decile)

RECODE x2ses (−1.7501 thru −0.8428 = 1) INTO sesdec.
RECODE x2ses (−0.8429 thru −0.5737 = 1) INTO sesdec.
RECODE x2ses (−0.5738 thru −0.3828 = 1) INTO sesdec.
RECODE x2ses (−0.3829 thru −0.1792 = 1) INTO sesdec.
RECODE x2ses (−0.1793 thru 0.0332 = 1) INTO sesdec.
RECODE x2ses (0.0333 thru 0.2582 = 1) INTO sesdec.
RECODE x2ses (0.2583 thru 0.4967 = 1) INTO sesdec.
RECODE x2ses (0.4968 thru 0.7782 = 1) INTO sesdec.
RECODE x2ses (0.7783 thru 1.1324 = 1) INTO sesdec.
RECODE x2ses (1.1325 thru 2.2824 = 1) INTO sesdec.

sesdec

		Frequency	Percent	Valid Percent	Cumulative Percent
	1.00	2013	10.0	10.0	10.0
	2.00	2013	10.0	10.0	20.0
	3.00	2016	10.0	10.0	30.0
	4.00	2011	10.0	10.0	40.0
	5.00	2012	10.0	10.0	50.0
Valid	6.00	2013	10.0	10.0	60.0
	7.00	2013	10.0	10.0	70.0
	8.00	2014	10.0	10.0	80.0
	9.00	2012	10.0	10.0	90.0
	10.00	2016	10.0	10.0	100.0
	Total	20133	100.0	100.0	

SES DECILE DUMMIES
(one dummy variable for each category of SES decile)

RECODE x2ses (−1.7501 thru −0.8428 = 1) (else=0) INTO sesd01.
RECODE x2ses (−0.8427 thru −0.5737 = 1) (else=0) INTO sesd02.
RECODE x2ses (−0.5736 thru −0.3828 = 1) (else=0) INTO sesd03.
RECODE x2ses (−0.3827 thru −0.1792 = 1) (else=0) INTO sesd04.
RECODE x2ses (−0.1791 thru 0.0332 = 1) (else=0) INTO sesd05.
RECODE x2ses (0.0333 thru 0.2582 = 1) (else=0) INTO sesd06.
RECODE x2ses (0.2583 thru 0.4967 = 1) (else=0) INTO sesd07.
RECODE x2ses (0.4968 thru 0.7782 = 1) (else=0) INTO sesd08.
RECODE x2ses (0.7783 thru 1.1324 = 1) (else=0) INTO sesd09.
RECODE x2ses (1.1325 thru 2.2824 = 1) (else=0) INTO sesd10.

sesdec * sesd10 Crosstabulation

Count

		sesd10		Total
		.00	1.00	
sesdec	1.00	2013	0	2013
	2.00	2013	0	2013
	3.00	2016	0	2016
	4.00	2011	0	2011
	5.00	2012	0	2012
	6.00	2013	0	2013
	7.00	2013	0	2013
	8.00	2014	0	2014
	9.00	2012	0	2012
	10.00	0	2016	2016
Total		18117	2016	20133

(Note: there are similar tables possible for the other decile variables)

Chapter 4. Key Regression Modeling Concepts

BIOLOGICAL PARENT/STEPPARENT+SINGLE
(two categories: biological parent/stepparent or single parent, other)

COMPUTE stepsing = step+single.

single * step Crosstabulation

Count

		step		Total
		.00	1.00	
single	.00	13207	2812	16019
	1.00	4114	0	4114
Total		17321	2812	20133

stepsing

		Frequency	Percent	Valid Percent	Cumulative Percent
Valid	.00	13207	65.6	65.6	65.6
	1.00	6926	34.4	34.4	100.0
	Total	20133	100.0	100.0	

ANSWERS TO SELECTED PROBLEMS

BIOLOGICAL PARENT/STEPPARENT+ OTHER FAMILY
COMPUTE stepfamoth=step+famoth.

SINGLE+ OTHER FAMILY
COMPUTE singlefamoth=single+famoth.

famoth * step Crosstabulation

Count

		step		Total
		.00	1.00	
famoth	.00	15557	2812	18369
	1.00	1764	0	1764
Total		17321	2812	20133

stepfamoth

		Frequency	Percent	Valid Percent	Cumulative Percent
Valid	.00	15557	77.3	77.3	77.3
	1.00	4576	22.7	22.7	100.0
	Total	20133	100.0	100.0	

(Note: there is a similar table possible for singlefamoth)

Chapter 5. Control Modeling

PARENTAL EDUCATION DUMMIES (FOUR CATEGORIES)

(one dummy variable for each of four categories: high school or less, 2 year degree, 4 year degree, graduate degree)

RECODE x2paredu (1 thru 2=1) (else=0) INTO hsorless.
RECODE x2paredu (3 thru 4=1) (else=0) INTO twoyr.
RECODE x2paredu (5=1) (else=0) INTO fouryr.
RECODE x2paredu (6 thru 7=1) (else=0) INTO grad.

x2paredu * twoyr Crosstabulation

Count

		twoyr		Total
		.00	1.00	
	1	1028	0	1028
	2	6112	0	6112
	3	0	960	960
x2paredu	4	0	3211	3211
	5	5002	0	5002
	6	2549	0	2549
	7	1271	0	1271
Total		15962	4171	20133

(Note: there are similar tables possible for hsorless, fouryr, and grad)

PARENTAL EDUCATION (FOUR CATEGORIES)
(four categories: <high school, high school only, 2 year degree, 4 year degree, MA/MS, PhD/MD)

RECODE x2paredu (1 thru 2=1) (3 thru 4=2) (5=3)
(6 thru 7=4) INTO paredfour.

x2paredu * paredfour Crosstabulation

Count

		paredfour				Total
		1.00	2.00	3.00	4.00	
x2paredu	1	1028	0	0	0	1028
	2	6112	0	0	0	6112
	3	0	960	0	0	960
	4	0	3211	0	0	3211
	5	0	0	5002	0	5002
	6	0	0	0	2549	2549
	7	0	0	0	1271	1271
Total		7140	4171	5002	3820	20133

FAMILY INCOME ($10,000 units)
(recode of categorical family income variable so that the value for category is the mid-point of family income category expressed in $10,000 units; the midpoint of the open-ended category was estimated by using a Pareto estimation)

RECODE x2famincome
(1=.75) (2=2.5) (3=4.5) (4=6.5) (5=8.5) (6=10.5) (7=12.5) (8=14.5) (9=16.5) (10=18.5)
(11=20.5) (12=22.5) (13=49.5) INTO faminc.

faminc

		Frequency	Percent	Valid Percent	Cumulative Percent
Valid	.75	1970	9.8	9.8	9.8
	2.50	3452	17.1	17.1	26.9
	4.50	3471	17.2	17.2	44.2
	6.50	2869	14.3	14.3	58.4
	8.50	2243	11.1	11.1	69.6
	10.50	1744	8.7	8.7	78.2
	12.50	1265	6.3	6.3	84.5
	14.50	849	4.2	4.2	88.7
	16.50	457	2.3	2.3	91.0
	18.50	319	1.6	1.6	92.6
	20.50	425	2.1	2.1	94.7
	22.50	168	.8	.8	95.5
	49.50	901	4.5	4.5	100.0
	Total	20133	100.0	100.0	

FAMILY INCOME (FOUR CATEGORIES)
(four categories: 0–$35,000, $36,000–$75,000, $76,000–$115,000, $116,000+)

RECODE x2famincome (1=1) (2=1) (3=2) (4=2) (5=3) (6=3) (else=4) INTO faminc4.

x2famincome * faminc4 Crosstabulation

Count

		faminc4				Total
		1.00	2.00	3.00	4.00	
x2famincome	1	1970	0	0	0	1970
	2	3452	0	0	0	3452
	3	0	3471	0	0	3471
	4	0	2869	0	0	2869
	5	0	0	2243	0	2243
	6	0	0	1744	0	1744
	7	0	0	0	1265	1265
	8	0	0	0	849	849
	9	0	0	0	457	457
	10	0	0	0	319	319
	11	0	0	0	425	425
	12	0	0	0	168	168
	13	0	0	0	901	901
Total		5422	6340	3987	4384	20133

WHITE
(two categories: white, not white)

RECODE x2race (8=1)(else=0) INTO white.

BLACK
(two categories: black, not black)

RECODE x2race (3=1)(else=0) INTO black.

OTHER RACE/ETHNCITY
(two categories: other race/ethnicity, not other race/ethnicity)

RECODE x2race (1 thru 2=1)(4 thru 7=1)(else=0) INTO othrace.

x2race * othrace Crosstabulation

Count

		othrace		Total
		.00	1.00	
x2race	1	0	129	129
	2	0	1641	1641
	3	2066	0	2066
	4	0	168	168
	5	0	3028	3028
	6	0	1711	1711
	7	0	96	96
	8	11294	0	11294
Total		13360	6773	20133

(Note: there are similar tables possible for white and black)

RACE/ETHNICITY (THREE CATEGORIES)
(three categories: white, black, other race/ethnicity)

RECODE x2race (8=1) (3=2) (else=3) INTO raceeth.

x2race * raceeth Crosstabulation

Count

		raceeth			Total
		1.00	2.00	3.00	
x2race	1	0	0	129	129
	2	0	0	1641	1641
	3	0	2066	0	2066
	4	0	0	168	168
	5	0	0	3028	3028
	6	0	0	1711	1711
	7	0	0	96	96
	8	11294	0	0	11294
Total		11294	2066	6773	20133

Chapter 6. Modeling Interactions

PARENT COLLEGE GRADUATE
(two categories: parent college graduate, parent not college graduate)

RECODE x2paredu (1 thru 4=0) (5 thru 7=1) INTO parcoll.

x2paredu * parcoll Crosstabulation

Count

		parcoll		Total
		.00	1.00	
x2paredu	1	1028	0	1028
	2	6112	0	6112
	3	960	0	960
	4	3211	0	3211
	5	0	5002	5002
	6	0	2549	2549
	7	0	1271	1271
Total		11311	8822	20133

PARENT NOT COLLEGE GRADUATE
(two categories: parent not college graduate, parent college graduate)

RECODE x2paredu (1 thru 4=1) (5 thru 7=0) INTO nparcoll.

x2paredu * nparcoll Crosstabulation

Count

		nparcoll		Total
		.00	1.00	
x2paredu	1	0	1028	1028
	2	0	6112	6112
	3	0	960	960
	4	0	3211	3211
	5	5002	0	5002
	6	2549	0	2549
	7	1271	0	1271
Total		8822	11311	20133

PARENT COLLEGE GRADUATE * TWO BIOLOGICAL PARENTS
(categories: parent college graduate and two biological parents, other)

COMPUTE pcolltwo = parcoll * twopar.

PARENT COLLEGE GRADUATE * BIOLOGICAL PARENT/STEPPARENT
(categories: parent college graduate and biological parent/stepparent, other)

COMPUTE pcollstep = parcoll * step.

PARENT COLLEGE GRADUATE * SINGLE PARENT
(categories: parent college graduate and single parent, other)

COMPUTE pcollsing = parcoll * single.

PARENT COLLEGE GRADUATE * OTHER FAMILY
(categories: parent college graduate and other family, other)

COMPUTE pcollfoth = parcoll * famoth.

parcoll * single Crosstabulation

Count

		single		Total
		.00	1.00	
parcoll	.00	8484	2827	11311
	1.00	7535	1287	8822
Total		16019	4114	20133

pcollsing

		Frequency	Percent	Valid Percent	Cumulative Percent
Valid	.00	18846	93.6	93.6	93.6
	1.00	1287	6.4	6.4	100.0
	Total	20133	100.0	100.0	

(Note: there are similar tables possible for the other three interaction variables)

PARENT NOT COLLEGE GRADUATE * TWO BIOLOGICAL PARENTS
(categories: parent not college graduate and two biological parents, other)

COMPUTE npcolltwo = nparcoll * twopar.

PARENT NOT COLLEGE GRADUATE * BIOLOGICAL PARENT/STEPPARENT
(categories: parent not college graduate and biological parent/stepparent, other)

COMPUTE npcollstep = nparcoll * step.

PARENT NOT COLLEGE GRADUATE * SINGLE PARENT
(categories: parent not college graduate and single parent, other)

COMPUTE npcollsing = nparcoll * single.

PARENT NOT COLLEGE GRADUATE * OTHER FAMILY
(categories: parent not college graduate and other family, other)

COMPUTE npcollfoth = nparcoll * famoth.

nparcoll * single Crosstabulation

Count

		single		Total
		.00	1.00	
nparcoll	.00	7535	1287	8822
	1.00	8484	2827	11311
Total		16019	4114	20133

npcollsing

		Frequency	Percent	Valid Percent	Cumulative Percent
Valid	.00	17306	86.0	86.0	86.0
	1.00	2827	14.0	14.0	100.0
	Total	20133	100.0	100.0	

(Note: there are similar tables possible for the other three interaction variables)

SES QUARTILE
(values: first quartile, second quartile, third quartile, fourth quartile)

RECODE x2ses
(Lowest thru −.480000=1)
(−.479999 thru .030000=2)
(.030001 thru .630000=3)
(.630001 thru Highest=4) INTO sesq4.

sesq4

		Frequency	Percent	Valid Percent	Cumulative Percent
Valid	1.00	4942	24.5	24.5	24.5
	2.00	5098	25.3	25.3	49.9
	3.00	5052	25.1	25.1	75.0
	4.00	5041	25.0	25.0	100.0
	Total	20133	100.0	100.0	

TWO BIOLOGICAL PARENTS * SES QUARTILE
(values: SES quartile values for those in two biological parents only)

COMPUTE twoparses = twopar * sesq4.

BIOLOGICAL PARENT/STEPPARENT * SES QUARTILE
(values: SES quartile values for those in biological parent/stepparent only)

COMPUTE stepses = step * sesq4.

SINGLE PARENT * SES QUARTILE
(values: SES quartile values for those in single only)

COMPUTE singleses = single * sesq4.

OTHER FAMILY * SES QUARTILE
(values: SES quartile values for those in other family only)

COMPUTE famothses = famoth * sesq4.

sesq4 * single Crosstabulation

Count

		single		Total
		.00	1.00	
sesq4	1.00	3338	1604	4942
	2.00	4022	1076	5098
	3.00	4111	941	5052
	4.00	4548	493	5041
Total		16019	4114	20133

singleses

		Frequency	Percent	Valid Percent	Cumulative Percent
Valid	.00	16019	79.6	79.6	79.6
	1.00	1604	8.0	8.0	87.5
	2.00	1076	5.3	5.3	92.9
	3.00	941	4.7	4.7	97.6
	4.00	493	2.4	2.4	100.0
	Total	20133	100.0	100.0	

(Note: there are similar tables possible for the other three interaction variables)

VARIABLE FOR SUM OF INTERACTION VARIABLES

COMPUTE intsum = stepses + singleses + famothses.

intsum * stepses Crosstabulation

Count

		stepses					Total
		.00	1.00	2.00	3.00	4.00	
intsum	.00	11443	0	0	0	0	11443
	1.00	2265	649	0	0	0	2914
	2.00	1589	0	818	0	0	2407
	3.00	1312	0	0	783	0	2095
	4.00	712	0	0	0	562	1274
Total		17321	649	818	783	562	20133

intsum * singleses Crosstabulation

Count

		singleses					Total
		.00	1.00	2.00	3.00	4.00	
intsum	.00	11443	0	0	0	0	11443
	1.00	1310	1604	0	0	0	2914
	2.00	1331	0	1076	0	0	2407
	3.00	1154	0	0	941	0	2095
	4.00	781	0	0	0	493	1274
Total		16019	1604	1076	941	493	20133

intsum * famothses Crosstabulation

Count

		famothses					Total
		.00	1.00	2.00	3.00	4.00	
intsum	.00	11443	0	0	0	0	11443
	1.00	2253	661	0	0	0	2914
	2.00	1894	0	513	0	0	2407
	3.00	1724	0	0	371	0	2095
	4.00	1055	0	0	0	219	1274
Total		18369	661	513	371	219	20133

WHITE * PARENT COLLEGE GRADUATE
(categories: white and parent college graduate, other)

COMPUTE wparcoll = white * parcoll.

BLACK * PARENT COLLEGE GRADUATE
(categories: black and parent college graduate, other)

COMPUTE bparcoll = black * parcoll.

OTHER RACE/ETHNICITY * PARENT COLLEGE GRADUATE
(categories: other race/ethncity and parent college graduate, other)

COMPUTE oparcoll = othrace * parcoll.

parcoll * white Crosstabulation

Count

		white		Total
		.00	1.00	
parcoll	.00	5479	5832	11311
	1.00	3360	5462	8822
Total		8839	11294	20133

wparcoll

		Frequency	Percent	Valid Percent	Cumulative Percent
Valid	.00	14671	72.9	72.9	72.9
	1.00	5462	27.1	27.1	100.0
	Total	20133	100.0	100.0	

(Note: there are similar tables possible for the other two interaction variables)

FEMALE
(two categories: female, not female)

RECODE x2sex (2=1)(1=0) INTO female.

x2sex * female Crosstabulation

Count

		female		Total
		0	1	
x2sex	1	10176	0	10176
	2	0	9957	9957
Total		10176	9957	20133

WHITE * FEMALE
(categories: white and female, other)

COMPUTE wfemale = white * female.

BLACK * FEMALE
(categories: black and female, other)

COMPUTE bfemale = black * female.

OTHER RACE/ETHNICITY * FEMALE
(categories: other race/ethncity and female, other)

COMPUTE ofemale = othrace * female.

female * black Crosstabulation

Count

		black		Total
		.00	1.00	
female	0	9113	1063	10176
	1	8954	1003	9957
Total		18067	2066	20133

bfemale

		Frequency	Percent	Valid Percent	Cumulative Percent
Valid	.00	19130	95.0	95.0	95.0
	1.00	1003	5.0	5.0	100.0
	Total	20133	100.0	100.0	

(Note: there are similar tables possible for the other two interaction variables)

PARENT COLLEGE GRADUATE * FEMALE
(categories: parent college graduate and female, other)

COMPUTE femcoll = female * parcoll.

parcoll * female Crosstabulation

Count

		female		Total
		0	1	
parcoll	.00	5697	5614	11311
	1.00	4479	4343	8822
Total		10176	9957	20133

femcoll

		Frequency	Percent	Valid Percent	Cumulative Percent
Valid	.00	15790	78.4	78.4	78.4
	1.00	4343	21.6	21.6	100.0
	Total	20133	100.0	100.0	

WHITE * PARENT COLLEGE GRADUATE * FEMALE
(categories: white and parent college graduate and female, other)

COMPUTE wfemcoll = white * female * parcoll.

BLACK * PARENT COLLEGE GRADUATE * FEMALE
(categories: black and parent college graduate and female, other)

COMPUTE bfemcoll = black * female * parcoll.

OTHER RACE/ETHNICITY * PARENT COLLEGE GRADUATE * FEMALE
(categories: other race/ethnicity and parent college graduate and female, other)

COMPUTE ofemcoll = othrace * female * parcoll.

black * femcoll Crosstabulation

Count

		femcoll		Total
		.00	1.00	
black	.00	14057	4010	18067
	1.00	1733	333	2066
Total		15790	4343	20133

bfemcoll

		Frequency	Percent	Valid Percent	Cumulative Percent
Valid	.00	19800	98.3	98.3	98.3
	1.00	333	1.7	1.7	100.0
	Total	20133	100.0	100.0	

(Note: there are similar tables possible for the other two interaction variables)

WHITE * PARENT NOT COLLEGE GRADUATE * FEMALE
(categories: white and parent not college graduate and female, other)

COMPUTE wfemncoll = white * female * nparcoll.

BLACK * PARENT NOT COLLEGE GRADUATE * FEMALE
(categories: black and parent not college graduate and female, other)

COMPUTE bfemncoll = black * female * nparcoll.

OTHER RACE/ETHNICITY * PARENT NOT COLLEGE GRADUATE * FEMALE
(categories: other race/ethnicity and parent not college graduate and female, other)

COMPUTE ofemncoll = othrace * female * nparcoll.

nparcoll * bfemale Crosstabulation

Count

		bfemale		Total
		.00	1.00	
nparcoll	.00	8489	333	8822
	1.00	10641	670	11311
Total		19130	1003	20133

bfemncoll

		Frequency	Percent	Valid Percent	Cumulative Percent
Valid	.00	19463	96.7	96.7	96.7
	1.00	670	3.3	3.3	100.0
	Total	20133	100.0	100.0	

(Note: there are similar tables possible for the other two interaction variables)

WHITE * FAMILY INCOME
(values: family income values for whites only)

COMPUTE wfaminc = white * faminc.

BLACK * FAMILY INCOME
(values: family income values for blacks only)

COMPUTE bfaminc = black * faminc.

OTHER RACE/ETHNICITY * FAMILY INCOME
(values: family income values for other race/ethnicity only)

COMPUTE ofaminc = othrace * faminc.

faminc * black Crosstabulation

Count

		black		Total
		.00	1.00	
faminc	.75	1622	348	1970
	2.50	3001	451	3452
	4.50	3050	421	3471
	6.50	2577	292	2869
	8.50	2084	159	2243
	10.50	1613	131	1744
	12.50	1182	83	1265
	14.50	799	50	849
	16.50	426	31	457
	18.50	292	27	319
	20.50	397	28	425
	22.50	162	6	168
	49.50	862	39	901
Total		18067	2066	20133

bfaminc

		Frequency	Percent	Valid Percent	Cumulative Percent
Valid	.00	18067	89.7	89.7	89.7
	.75	348	1.7	1.7	91.5
	2.50	451	2.2	2.2	93.7
	4.50	421	2.1	2.1	95.8
	6.50	292	1.5	1.5	97.2
	8.50	159	.8	.8	98.0
	10.50	131	.7	.7	98.7
	12.50	83	.4	.4	99.1
	14.50	50	.2	.2	99.3
	16.50	31	.2	.2	99.5
	18.50	27	.1	.1	99.6
	20.50	28	.1	.1	99.8
	22.50	6	.0	.0	99.8
	49.50	39	.2	.2	100.0
	Total	20133	100.0	100.0	

(Note: there are similar tables possible for the other two interaction variables)

BLACK * PARENT NOT COLLEGE GRADUATE
(categories: black and parent not college graduate, other)

COMPUTE bnpcoll = black * nparcoll.

OTHER RACE/ETHNICITY * PARENT NOT COLLEGE GRADUATE
(categories: other race/ethncity and parent not college graduate, other)

COMPUTE onpcoll = othrace * nparcoll.

nparcoll * black Crosstabulation

Count

		black		Total
		.00	1.00	
nparcoll	.00	8098	724	8822
	1.00	9969	1342	11311
Total		18067	2066	20133

bnpcoll

		Frequency	Percent	Valid Percent	Cumulative Percent
Valid	.00	18791	93.3	93.3	93.3
	1.00	1342	6.7	6.7	100.0
	Total	20133	100.0	100.0	

(Note: there are similar tables possible for the other interaction variable)

ANSWERS TO SELECTED PROBLEMS

TWO BIOLOGICAL PARENTS * FAMILY INCOME
(values: family income values for those in two biological parents only)

COMPUTE twoinc = twopar * faminc.

BIOLOGICAL PARENT/STEPPARENT * FAMILY INCOME
(values: family income values for those in biological parent/stepparent only)

COMPUTE stepinc = step * faminc.

SINGLE PARENT * FAMILY INCOME
(values: family income values for those in single only)

COMPUTE singleinc = single * faminc.

OTHER FAMILY * FAMILY INCOME
(values: family income values for those in other family only)

COMPUTE famothinc = famoth * faminc.

faminc * step Crosstabulation

Count

		step .00	step 1.00	Total
faminc	.75	1720	250	1970
	2.50	2968	484	3452
	4.50	2919	552	3471
	6.50	2414	455	2869
	8.50	1922	321	2243
	10.50	1522	222	1744
	12.50	1097	168	1265
	14.50	737	112	849
	16.50	407	50	457
	18.50	282	37	319
	20.50	369	56	425
	22.50	152	16	168
	49.50	812	89	901
Total		17321	2812	20133

stepinc

		Frequency	Percent	Valid Percent	Cumulative Percent
Valid	.0000	17321	86.0	86.0	86.0
	.7500	250	1.2	1.2	87.3
	2.5000	484	2.4	2.4	89.7
	4.5000	552	2.7	2.7	92.4
	6.5000	455	2.3	2.3	94.7
	8.5000	321	1.6	1.6	96.3
	10.5000	222	1.1	1.1	97.4
	12.5000	168	.8	.8	98.2
	14.5000	112	.6	.6	98.8
	16.5000	50	.2	.2	99.0
	18.5000	37	.2	.2	99.2
	20.5000	56	.3	.3	99.5
	22.5000	16	.1	.1	99.6
	49.5000	89	.4	.4	100.0
	Total	20133	100.0	100.0	

(Note: there are similar tables possible for the other two interaction variables)

WHITE * TWO BIOLOGICAL PARENTS
(categories: white and two biological parents, other)

COMPUTE wtwopar = white * twopar.

BLACK * TWO BIOLOGICAL PARENTS
(categories: black and two biological parents, other)

COMPUTE btwopar = black * twopar.

OTHER RACE/ETHNICITY * TWO BIOLOGICAL PARENTS
(categories: other race/ethncity and two biological parents, other)

COMPUTE otwopar = othrace * twopar.

twopar * white Crosstabulation

Count

		white		Total
		.00	1.00	
twopar	.00	4131	4559	8690
	1.00	4708	6735	11443
Total		8839	11294	20133

wtwopar

		Frequency	Percent	Valid Percent	Cumulative Percent
Valid	.0000	13398	66.5	66.5	66.5
	1.0000	6735	33.5	33.5	100.0
	Total	20133	100.0	100.0	

(Note: there are similar tables possible for the other two interaction variables)

Chapter 7. Modeling Linearity With Splines

SES QUARTILE DUMMIES
(one dummy for each of four categories: first quartile, second quartile, third quartile, fourth quartile)

RECODE x2ses (Lowest thru −.480000=1) (else=0) INTO sesq4_1.
RECODE x2ses (−.479999 thru .030000=1) (else=0) INTO sesq4_2.
RECODE x2ses (.030001 thru .630000=1) (else=0) INTO sesq4_3.
RECODE x2ses (.630001 thru Highest=1) (else=0) INTO sesq4_4.

sesq4_1 * sesq4_2 Crosstabulation

Count

		sesq4_2		Total
		.00	1.00	
sesq4_1	.00	10093	5098	15191
	1.00	4942	0	4942
Total		15035	5098	20133

sesq4_1 * sesq4_3 Crosstabulation

Count

		sesq4_3		Total
		.00	1.00	
sesq4_1	.00	10139	5052	15191
	1.00	4942	0	4942
Total		15081	5052	20133

sesq4_1 * sesq4_4 Crosstabulation

Count

		sesq4_4		Total
		.00	1.00	
sesq4_1	.00	10150	5041	15191
	1.00	4942	0	4942
Total		15092	5041	20133

THIRD QUARTILE JUMP VARIABLE

COMPUTE jsesq4_3 = sesq4_3+sesq4_4.

jsesq4_3 * sesq4_3 Crosstabulation

Count

		sesq4_3		Total
		.00	1.00	
jsesq4_3	.00	10040	0	10040
	1.00	5041	5052	10093
Total		15081	5052	20133

jsesq43 * sesq4_4 Crosstabulation

Count

		sesq4_4		Total
		.00	1.00	
jsesq4_3	.00	10040	0	10040
	1.00	5052	5041	10093
Total		15092	5041	20133

SECOND QUARTILE JUMP VARIABLE

COMPUTE jsesq4_2 = sesq4_2+sesq4_3 +sesq4_4.
or
COMPUTE jsesq4_2 = sesq4_2 +j sesq4_3.

jsesq4_2 * sesq4_2 Crosstabulation

Count

		sesq4_2		Total
		.00	1.00	
jsesq4_2	.00	4942	0	4942
	1.00	10093	5098	15191
Total		15035	5098	20133

jsesq4_2 * jsesq4_3 Crosstabulation

Count

		jsesq4_3		Total
		.00	1.00	
	.00	4942	0	4942
jsesq4_2	1.00	5098	10093	15191
Total		10040	10093	20133

JUMP SUMMER

COMPUTE sesjsum = jsesq4_2+jsesq4_3 +sesq4_4.

sesjsum * jsesq4_2 Crosstabulation

Count

		jsesq4_2		Total
		.00	1.00	
sesjsum	.00	4942	0	4942
	1.00	0	5098	5098
	2.00	0	5052	5052
	3.00	0	5041	5041
Total		4942	15191	20133

sesjsum * jsesq4_3 Crosstabulation

Count

		jsesq4_3		Total
		.00	1.00	
sesjsum	.00	4942	0	4942
	1.00	5098	0	5098
	2.00	0	5052	5052
	3.00	0	5041	5041
Total		10040	10093	20133

sesjsum * sesq4_4 Crosstabulation

Count

		sesq4_4		Total
		.00	1.00	
sesjsum	.00	4942	0	4942
	1.00	5098	0	5098
	2.00	5052	0	5052
	3.00	0	5041	5041
Total		15092	5041	20133

X2SES-12 (12 INTEGER VALUES)

RECODE x2ses (−1.7501 thru −1.4167 = 1) INTO x2ses_12.
RECODE x2ses (−1.4166 thru −1.0835 = 2) INTO x2ses_12.
RECODE x2ses (−1.0834 thru −0.7502 = 3) INTO x2ses_12.
RECODE x2ses (−0.7501 thru −0.4167 = 4) INTO x2ses_12.
RECODE x2ses (−0.4166 thru −0.0834 = 5) INTO x2ses_12.
RECODE x2ses (−0.0833 thru 0.2499 = 6) INTO x2ses_12.
RECODE x2ses (0.2500 thru 0.5833 = 7) INTO x2ses_12.
RECODE x2ses (0.5834 thru 0.9167 = 8) INTO x2ses_12.
RECODE x2ses (0.9168 thru 1.2501 = 9) INTO x2ses_12.
RECODE x2ses (1.2502 thru 1.5835 =10) INTO x2ses_12.
RECODE x2ses (1.5836 thru 1.9169 =11) INTO x2ses_12.
RECODE x2ses (1.9170 thru 2.2824 =12) INTO x2ses_12.

x2ses_12

		Frequency	Percent	Valid Percent	Cumulative Percent
Valid	1.0000	188	.9	.9	.9
	2.0000	694	3.4	3.4	4.4
	3.0000	1695	8.4	8.4	12.8
	4.0000	3068	15.2	15.2	28.0
	5.0000	3339	16.6	16.6	44.6
	6.0000	3015	15.0	15.0	59.6
	7.0000	2765	13.7	13.7	73.3
	8.0000	2222	11.0	11.0	84.4
	9.0000	1671	8.3	8.3	92.7
	10.0000	1002	5.0	5.0	97.6
	11.0000	382	1.9	1.9	99.5
	12.0000	92	.5	.5	100.0
	Total	20133	100.0	100.0	

CREATION OF SPLINE VARIABLES FOR X2SES-12

RECODE x2ses_12 (1=1) (2 thru 12=2) (ELSE=0) INTO S_1_2.
RECODE x2ses_12 (3=1) (4 thru 12=2) (ELSE=0) INTO S_3_4.
RECODE x2ses_12 (5=1) (6 thru 12=2) (ELSE=0) INTO S_5_6.
RECODE x2ses_12 (7=1) (8 thru 12=2) (ELSE=0) INTO S_7_8.
RECODE x2ses_12 (9=1) (10 thru 12=2) (ELSE=0) INTO S_9_10.
RECODE x2ses_12 (11=1) (12 thru 12=2) (ELSE=0) INTO S_11_12.

COMPUTE D_9_10 = S_9_10 + S_11_12.
COMPUTE D_7_8 = S_7_8 + D_9_10.
COMPUTE D_5_6 = S_5_6 + D_7_8.
COMPUTE D_3_4 = S_3_4 + D_5_6.
COMPUTE D_1_2 = S_1_2 + D_3_4.
COMPUTE S_1_6 = S_1_2 + S_3_4 + S_5_6.
COMPUTE S_7_12 = S_7_8 + S_9_10 + S_11_12.
COMPUTE S_1_8 = S_1_2 + S_3_4 + S_5_6 + S_7_8.
COMPUTE S_9_12 = S_9_10 + S_11_12.

COMPUTE check1= S_1_2 + S_3_4 + S_5_6 + S_7_8 + S_9_10 + S_11_12.

check1

		Frequency	Percent	Valid Percent	Cumulative Percent
	1.00	188	.9	.9	.9
	2.00	694	3.4	3.4	4.4
	3.00	1695	8.4	8.4	12.8
	4.00	3068	15.2	15.2	28.0
	5.00	3339	16.6	16.6	44.6
	6.00	3015	15.0	15.0	59.6
Valid	7.00	2765	13.7	13.7	73.3
	8.00	2222	11.0	11.0	84.4
	9.00	1671	8.3	8.3	92.7
	10.00	1002	5.0	5.0	97.6
	11.00	382	1.9	1.9	99.5
	12.00	92	.5	.5	100.0
	Total	20133	100.0	100.0	

CREATION OF SPLINE VARIABLES S FOR CONTINUOUS SES VARIABLE

COMPUTE S01=0.
COMPUTE S02=0.
COMPUTE S03=0.
COMPUTE S04=0.
COMPUTE S05=0.
COMPUTE S06=0.
IF −1.7501<=x2ses and x2ses<=−1.0835 S01=x2ses−(−1.7501).
IF x2SES>−1.0835 S01=.6666.
IF −1.0835<x2ses and x2ses<=−0.4167 S02=x2ses−(−1.0835).
IF x2SES>−0.4167 S02=.6668.
IF −0.4167<x2ses and x2ses<= 0.2499 S03=x2ses−(−0.4167).
IF x2SES> 0.2499 S03=.6666.
IF 0.2499<x2ses and x2ses<= 0.9167 S04=x2ses−(0.2499).
IF x2SES> 0.9167 S04=.6668.
IF 0.9167<x2ses and x2ses<= 1.5835 S05=x2ses−(0.9167).
IF x2SES> 1.5835 S05=.6668.
IF 1.5835<x2ses and x2ses<= 2.2824 S06=x2ses−(1.5835).

COMPUTE S_01_03 = S01 + S02 + S03.
COMPUTE S_04_06 = S04 + S05 + S06.

COMPUTE check2= S01+S02+S03+S04+S05+S06+(−1.7501).

Descriptive Statistics

	N	Minimum	Maximum	Mean	Std. Deviation
x2ses	20133	-1.75	2.28	.0908	.74541
check2	20133	-1.75	2.28	.0908	.74541
Valid N (listwise)	20133				

ANSWERS TO SELECTED PROBLEMS

COMPUTE S1=0.
COMPUTE S2=0.
COMPUTE S3=0.
COMPUTE S4=0.
IF -1.7501<=x2ses and x2ses<=−0.7501 S1=x2ses−(−1.7501).
IF x2SES>−.7501 S1=1.0000.

IF -0.7501<x2ses and x2ses<=0.2499 S2=x2ses−(−0.7501).
IF x2SES>0.2499 S2=1.0000.

IF 0.2499<x2ses and x2ses<= 1.2499 S3=x2ses−(0.2499).
IF x2SES> 1.2499 S3=1.0000.

IF 1.2499<x2ses and x2ses<= 2.2824 S4=x2ses−(1.2499).

COMPUTE S12 = S1 + S2.
COMPUTE S34 = S3 + S4.

COMPUTE check3= S1+S2+S3+S4+(−1.7501).

Descriptive Statistics

	N	Minimum	Maximum	Mean	Std. Deviation
x2ses	20133	-1.75	2.28	.0908	.74541
check3	20133	-1.75	2.28	.0908	.74541
Valid N (listwise)	20133				

Answers to Chapter Exercises

Chapter 2. Basic Statistical Procedures

1,2. The *N* for the newly constructed data set should be 20,133. You should be able to reproduce the following frequency distribution for X2CONTROL:

x2control

		Frequency	Percent	Valid Percent	Cumulative Percent
Valid	1	16797	83.4	83.4	83.4
	2	3336	16.6	16.6	100.0
	Total	20133	100.0	100.0	

3. Use the means procedure to produce the following table:

x2txmtscor

famstruct	Mean	N	Std. Deviation
1.00	52.8808	11443	10.08391
2.00	50.3221	2812	9.82744
3.00	50.1456	4114	9.88436
4.00	49.1048	1764	9.94861
Total	51.6337	20133	10.10163

For example, the difference in means between those living with a biological parent/stepparent and two biological parents is −2.56 and equals 50.3221 minus 52.8808.

4. Use the cross-tabulation procedure to produce the following table:

highmath * famstruct Crosstabulation

Count

		famstruct				Total
		1.00	2.00	3.00	4.00	
highmath	.00	8126	2232	3284	1466	15108
	1.00	3317	580	830	298	5025
Total		11443	2812	4114	1764	20133

The SPSS program (SPSS Corporation, Chicago, IL) does not produce the log odds, so the log odds need to be calculated by using the numbers from the frequency table. For example, the difference in log odds between those living with a biological parent/stepparent and two biological parents is −.45 and equals ln(580/2232) minus ln(3317/8126).

5. Use the means procedure to produce the following table:

x2ses

famstruct	Mean	N	Std. Deviation
1.00	.2597	11443	.74808
2.00	.0501	2812	.65978
3.00	-.2281	4114	.67459
4.00	-.1960	1764	.67075
Total	.0908	20133	.74541

The new table of differences should look like Table 2.12.

6. Use the cross-tabulation procedure to produce the following table:

private * famstruct Crosstabulation

Count

		famstruct				Total
		1.00	2.00	3.00	4.00	
private	.00	9032	2535	3637	1593	16797
	1.00	2411	277	477	171	3336
Total		11443	2812	4114	1764	20133

Calculate the log odds by using the frequencies from the table above. The new table of differences should look like Table 2.14.

Chapter 3. Regression Modeling Basics

1. Use the linear regression procedure with X2TXMTSCOR as the dependent variable to produce the coefficients for each model. Create the four models by leaving one of TWOPAR, STEP, SINGLE, and FAMOTH out of the model. The results for Model 1 are as follows:

Coefficients[a]

Model		Unstandardized Coefficients		Standardized Coefficients	t	Sig.
		B	Std. Error	Beta		
1	(Constant)	52.881	.093		565.902	.000
	step	-2.559	.210	-.088	-12.162	.000
	single	-2.735	.182	-.109	-15.053	.000
	famoth	-3.776	.256	-.106	-14.768	.000

a. Dependent Variable: x2txmtscor

2. Use the binary logistic regression procedure with HIGHMATH to produce the coefficients for each model. Create the four models by leaving one of TWOPAR, STEP, SINGLE, and FAMOTH out of the model. The results for Model 1 are as follows:

Variables in the Equation

		B	S.E.	Wald	df	Sig.	Exp(B)
Step 1[a]	step	-.452	.051	78.545	1	.000	.637
	single	-.479	.044	118.834	1	.000	.619
	famoth	-.697	.067	108.928	1	.000	.498
	Constant	-.896	.021	1891.064	1	.000	.408

a. Variable(s) entered on step 1: step, single, famoth.

3. Use the linear regression procedure with X2SES as the dependent variable to produce the coefficients for each model. Create the four models by leaving one of TWOPAR, STEP, SINGLE, and FAMOTH out of the model. The results for Model 1 are as follows:

Coefficients[a]

Model		Unstandardized Coefficients		Standardized Coefficients	t	Sig.
		B	Std. Error	Beta		
1	(Constant)	.260	.007		38.855	.000
	step	-.210	.015	-.097	-13.930	.000
	single	-.488	.013	-.264	-37.527	.000
	famoth	-.456	.018	-.173	-24.915	.000

a. Dependent Variable: x2ses

4. Use the binary logistic regression procedure with PRIVATE to produce the coefficients for each model. Create the four models by leaving one of TWOPAR, STEP, SINGLE, and FAMOTH out of the model. The results for Model 1 are as follows:

Variables in the Equation

		B	S.E.	Wald	df	Sig.	Exp(B)
Step 1[a]	step	-.893	.067	176.113	1	.000	.409
	single	-.711	.054	174.342	1	.000	.491
	famoth	-.911	.084	118.535	1	.000	.402
	Constant	-1.321	.023	3319.486	1	.000	.267

a. Variable(s) entered on step 1: step, single, famoth.

5. Use the binary logistic regression procedure with PRIVATE to produce the coefficients for each model. Create the model by including SESQ52, SESQ53, SESQ54, and SESQ55 in the model. The results and plot for the model are as follows:

Variables in the Equation

		B	S.E.	Wald	df	Sig.	Exp(B)
Step 1[a]	sesq52	.835	.123	46.200	1	.000	2.305
	sesq53	1.319	.116	129.131	1	.000	3.742
	sesq54	1.942	.111	307.547	1	.000	6.970
	sesq55	2.738	.108	645.851	1	.000	15.452
	Constant	-3.429	.104	1093.056	1	.000	.032

a. Variable(s) entered on step 1: sesq2, sesq3, sesq4, sesq5.

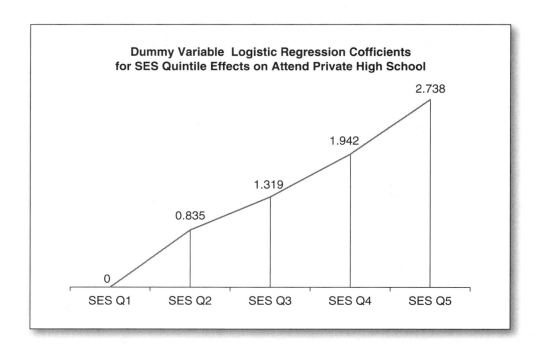

Chapter 4. Key Regression Modeling Concepts

1. Use the linear regression procedure with STEP, SINGLE, FAMOTH, and STEPSING as the independent variables and X2TXMTSCOR as the dependent variable to produce the coefficients for each model.

Coefficients[a]

Model		Unstandardized Coefficients		Standardized Coefficients	t	Sig.
		B	Std. Error	Beta		
1	(Constant)	52.881	.093		565.902	.000
	step	-2.559	.210	-.088	-12.162	.000
	single	-2.735	.182	-.109	-15.053	.000
	famoth	-3.776	.256	-.106	-14.768	.000

a. Dependent Variable: x2txmtscor

Coefficients[a]

Model		Unstandardized Coefficients		Standardized Coefficients	t	Sig.
		B	Std. Error	Beta		
1	(Constant)	52.881	.093		565.902	.000
	stepsing	-2.559	.210	-.120	-12.162	.000
	single	-.176	.245	-.007	-.722	.471
	famoth	-3.776	.256	-.106	-14.768	.000

a. Dependent Variable: x2txmtscor

Coefficients[a]

Model		Unstandardized Coefficients		Standardized Coefficients	t	Sig.
		B	Std. Error	Beta		
1	(Constant)	52.881	.093		565.908	.000
	stepsing	-2.664	.152	-.125	-17.503	.000
	famoth	-3.776	.256	-.106	-14.768	.000

a. Dependent Variable: x2txmtscor

2. Use the linear regression procedure with STEP, SINGLE, FAMOTH, and STEPFAMOTH, and SINGLEFAMOTH as the independent variables and X2TXMTSCOR as the dependent variable to produce the coefficients for the models.

Coefficients[a]

Model		Unstandardized Coefficients		Standardized Coefficients	t	Sig.
		B	Std. Error	Beta		
1	(Constant)	52.881	.093		565.902	.000
	stepfamoth	-2.559	.210	-.106	-12.162	.000
	single	-2.735	.182	-.109	-15.053	.000
	famoth	-1.217	.304	-.034	-4.009	.000

a. Dependent Variable: x2txmtscor

The coefficient for FAMOTH in the linear regression model above is significantly different from zero since the t value of -4.009 is greater than 1.96, which is the critical value at the .05 level of significance. The coefficient shows that those living in other families have lower test scores than those who live with a biological parent/stepparent.

Coefficients[a]

Model		Unstandardized Coefficients		Standardized Coefficients	t	Sig.
		B	Std. Error	Beta		
1	(Constant)	52.881	.093		565.902	.000
	sep	-2.559	.210	-.088	-12.162	.000
	singlefamoth	-2.735	.182	-.123	-15.053	.000
	famoth	-1.041	.284	-.029	-3.658	.000

a. Dependent Variable: x2txmtscor

The coefficient for FAMOTH in the linear regression model above is significantly different from zero since the t value of $|-3.658|$ is greater than 1.96. The coefficient shows that those living in other families have lower test scores than those who live with a single parent.

3. Use the binary logistic regression procedure with STEP, SINGLE, FAMOTH, STEPSINGLE, STEPFAMOTH, and SINGLEFAMOTH as the independent variables and PRIVATE as the dependent variable to produce the coefficients for the models.

Variables in the Equation

		B	S.E.	Wald	df	Sig.	Exp(B)
Step 1[a]	step	-.893	.067	176.113	1	.000	.409
	single	-.711	.054	174.342	1	.000	.491
	famoth	-.911	.084	118.535	1	.000	.402
	Constant	-1.321	.023	3319.486	1	.000	.267

a. Variable(s) entered on step 1: step, single, famoth.

The coefficients in the logistic regression model above represent the differences chances of attending private high school for those with a biological parent/stepparent, single parent, or in other family and those from those with two biological parents. Each coefficient is significantly different from zero.

Variables in the Equation

		B	S.E.	Wald	df	Sig.	Exp(B)
Step 1[a]	stepsing	-.893	.067	176.113	1	.000	.409
	single	.183	.080	5.226	1	.022	1.200
	famoth	-.911	.084	118.535	1	.000	.402
	Constant	-1.321	.023	3319.486	1	.000	.267

a. Variable(s) entered on step 1: stepsing, single, famoth.

The coefficient for SINGLE in the logistic regression model above shows that the coefficient for SINGLE is significantly less negative than the coefficient for STEP.

Variables in the Equation

		B	S.E.	Wald	df	Sig.	Exp(B)
Step 1ª	stepfamoth	-.893	.067	176.113	1	.000	.409
	single	-.711	.054	174.342	1	.000	.491
	famoth	-.018	.102	.030	1	.862	.982
	Constant	-1.321	.023	3319.486	1	.000	.267

a. Variable(s) entered on step 1: stepfamoth, single, famoth.

The coefficient for FAMOTH in the logistic regression model above shows that the coefficient for FAMOTH is not significantly different from the coefficient for STEP.

Variables in the Equation

		B	S.E.	Wald	df	Sig.	Exp(B)
Step 1ª	step	-.893	.067	176.113	1	.000	.409
	singlefamoth	-.711	.054	174.342	1	.000	.491
	famoth	-.200	.094	4.535	1	.033	.818
	Constant	-1.321	.023	3319.486	1	.000	.267

a. Variable(s) entered on step 1: step, singlefamoth, famoth.

The coefficient for FAMOTH in the logistic regression model above shows that the coefficient for FAMOTH is significantly more negative than the coefficient for SINGLE.

Chapter 5. Control Modeling

1. Estimate models using the linear regression procedure with X2TXMTSCOR as the dependent variable and the following independent variables:

model 1: PRIVATE
model 2: PRIVATE FAMINC
model 3: PRIVATE TWOYR FOURYR GRAD
model 4: PRIVATE STEP SINGLE FAMOTH
model 5: PRIVATE FAMINC TWOYR FOURYR GRAD STEP SINGLE FAMOTH

Coefficients[a]

Model		Unstandardized Coefficients		Standardized Coefficients	t	Sig.
		B	Std. Error	Beta		
1	(Constant)	50.795	.077		663.297	.000
	private	5.063	.188	.186	26.912	.000

a. Dependent Variable: x2txmtscor

Coefficients[a]

Model		Unstandardized Coefficients		Standardized Coefficients	t	Sig.
		B	Std. Error	Beta		
1	(Constant)	49.012	.092		531.887	.000
	private	3.141	.192	.116	16.333	.000
	x2txmtscor	.234	.007	.233	32.946	.000

a. Dependent Variable: x2txmtscor

Coefficients[a]

Model		Unstandardized Coefficients		Standardized Coefficients	t	Sig.
		B	Std. Error	Beta		
1	(Constant)	47.879	.112		428.937	.000
	private	2.665	.184	.098	14.482	.000
	twoyr	1.319	.183	.053	7.218	.000
	fouryr	5.571	.176	.238	31.737	.000
	grad	8.724	.193	.339	45.180	.000

a. Dependent Variable: x2txmtscor

Coefficients[a]

Model		Unstandardized Coefficients		Standardized Coefficients	t	Sig.
		B	Std. Error	Beta		
1	(Constant)	51.909	.100		517.565	.000
	private	4.614	.189	.170	24.466	.000
	step	-2.041	.208	-.070	-9.794	.000
	single	-2.298	.180	-.092	-12.770	.000
	famoth	-3.251	.253	-.091	-12.856	.000

a. Dependent Variable: x2txmtscor

Coefficients^a

Model		Unstandardized Coefficients		Standardized Coefficients	t	Sig.
		B	Std. Error	Beta		
1	(Constant)	47.917	.138		346.111	.000
	private	1.777	.188	.065	9.472	.000
	faminc	.125	.007	.124	17.044	.000
	twoyr	1.139	.181	.046	6.282	.000
	fouryr	4.794	.178	.205	26.880	.000
	grad	7.340	.204	.285	35.955	.000
	step	-1.326	.197	-.046	-6.738	.000
	single	-.824	.172	-.033	-4.791	.000
	famoth	-1.877	.240	-.053	-7.828	.000

a. Dependent Variable: x2txmtscor

2. The bivariate analysis is as follows:

x2txmtscor * twopar

x2txmtscor

twopar	Mean	N	Std. Deviation
.00	49.9914	8690	9.88837
1.00	52.8808	11443	10.08391
Total	51.6337	20133	10.10163

x2txmtscor * faminc

x2txmtscor

faminc	Mean	N	Std. Deviation
.75	46.9424	1970	9.38680
2.50	48.2932	3452	9.46419
4.50	50.0767	3471	9.66876
6.50	51.4083	2869	9.41043
8.50	52.6537	2243	9.38915
10.50	53.9544	1744	10.09528
12.50	54.6484	1265	9.67861
14.50	55.6075	849	9.54603
16.50	55.9684	457	10.04561
18.50	56.7164	319	9.73283
20.50	57.0309	425	9.55787
22.50	56.5382	168	9.72732
49.50	58.9381	901	9.62213
Total	51.6337	20133	10.10163

x2txmtscor * paredfour

x2txmtscor

paredfour	Mean	N	Std. Deviation
1.00	48.0664	7140	9.14784
2.00	49.4768	4171	9.08100
3.00	54.0894	5002	9.50840
4.00	57.4408	3820	10.12683
Total	51.6337	20133	10.10163

x2txmtscor * private

x2txmtscor

private	Mean	N	Std. Deviation
.00	50.7948	16797	10.11718
1.00	55.8577	3336	8.89369
Total	51.6337	20133	10.10163

faminc * twopar Crosstabulation

% within twopar

		twopar		Total
		.00	1.00	
faminc	.75	14.0%	6.6%	9.8%
	2.50	22.2%	13.3%	17.1%
	4.50	19.1%	15.9%	17.2%
	6.50	13.3%	14.9%	14.3%
	8.50	10.0%	12.0%	11.1%
	10.50	6.4%	10.4%	8.7%
	12.50	4.6%	7.6%	6.3%
	14.50	3.2%	5.0%	4.2%
	16.50	1.7%	2.7%	2.3%
	18.50	1.2%	1.9%	1.6%
	20.50	1.5%	2.6%	2.1%
	22.50	0.4%	1.1%	0.8%
	49.50	2.3%	6.1%	4.5%
Total		100.0%	100.0%	100.0%

paredfour * twopar Crosstabulation

% within twopar

		twopar		Total
		.0000	1.0000	
paredfour	1.0000	43.1%	29.7%	35.5%
	2.0000	23.2%	18.8%	20.7%
	3.0000	20.6%	28.1%	24.8%
	4.0000	13.0%	23.5%	19.0%
Total		100.0%	100.0%	100.0%

private * twopar Crosstabulation

% within twopar

		twopar		Total
		.0000	1.0000	
private	.0000	89.4%	78.9%	83.4%
	1.0000	10.6%	21.1%	16.6%
Total		100.0%	100.0%	100.0%

Those living with two parents have higher math scores than those not living with two parents. Those with higher family income have higher math scores than those with lower family income. Those whose parents have more education have higher math scores than those whose parents have less education. Those who attend private high school have higher math scores than those who do not attend private high school.

Those living with two parents have higher family income than those not living with two parents. Those living with two parents have higher parental education than those not living with two parents. Those living with two parents are more likely to attend private high school than those not living with two parents.

Estimate models by using the linear regression procedure with X2TXMTSCOR as the dependent variable and the following independent variables:

model 1: TWOPAR
model 2: TWOPAR FAMINC
model 3: TWOPAR TWOYR FOURYR GRAD
model 4: TWOPAR PRIVATE
model 5: TWOPAR FAMINC TWOYR FOURYR GRAD PRIVATE

Coefficients[a]

Model		Unstandardized Coefficients		Standardized Coefficients	t	Sig.
		B	Std. Error	Beta		
1	(Constant)	49.991	.107		466.022	.000
	twopar	2.889	.142	.142	20.307	.000

a. Dependent Variable: x2txmtscor

Coefficients[a]

Model		Unstandardized Coefficients		Standardized Coefficients	t	Sig.
		B	Std. Error	Beta		
1	(Constant)	48.223	.115		421.131	.000
	twopar	2.016	.140	.099	14.421	.000
	faminc	.253	.007	.251	36.680	.000

a. Dependent Variable: x2txmtscor

Coefficients[a]

Model		Unstandardized Coefficients		Standardized Coefficients	t	Sig.
		B	Std. Error	Beta		
1	(Constant)	47.315	.128		368.217	.000
	twopar	1.580	.136	.077	11.628	.000
	twoyr	1.346	.183	.054	7.356	.000
	fouryr	5.760	.175	.246	33.000	.000
	grad	9.014	.191	.350	47.276	.000

a. Dependent Variable: x2txmtscor

Coefficients[a]

Model		Unstandardized Coefficients		Standardized Coefficients	t	Sig.
		B	Std. Error	Beta		
1	(Constant)	49.500	.108		460.023	.000
	twopar	2.408	.142	.118	17.007	.000
	private	4.617	.189	.170	24.480	.000

a. Dependent Variable: x2txmtscor

Coefficients[a]

Model		Unstandardized Coefficients		Standardized Coefficients	t	Sig.
		B	Std. Error	Beta		
1	(Constant)	46.724	.131		357.942	.000
	twopar	1.201	.135	.059	8.867	.000
	faminc	.125	.007	.124	17.019	.000
	twoyr	1.124	.181	.045	6.203	.000
	fouryr	4.786	.178	.205	26.853	.000
	grad	7.323	.204	.284	35.888	.000
	private	1.792	.188	.066	9.557	.000

a. Dependent Variable: x2txmtscor

Controlling for family income explained 30% of the two-biological-parent effect found in the smallest model ($(2.016 − 2.889) × 100/2.889$) while controlling for parental education explained 45% ($(1.580 − 2.889) × 100/2.889$) and controlling for private high school explained 17% ($(2.408 − 2.889) × 100/2.889$). Controlling for all three variables explained 58% ($(1.201 − 2.889) × 100/2.889$).

3. The bivariate analysis is as follows:

twopar * highmath Crosstabulation

% within twopar

		highmath		Total
		.00	1.00	
twopar	.00	80.3%	19.7%	100.0%
	1.00	71.0%	29.0%	100.0%
Total		75.0%	25.0%	100.0%

faminc * highmath Crosstabulation

% within faminc_10

		highmath		Total
		.00	1.00	
faminc	.75	88.6%	11.4%	100.0%
	2.50	85.8%	14.2%	100.0%
	4.50	80.3%	19.7%	100.0%
	6.50	77.8%	22.2%	100.0%
	8.50	72.9%	27.1%	100.0%
	10.50	68.1%	31.9%	100.0%
	12.50	64.7%	35.3%	100.0%
	14.50	61.7%	38.3%	100.0%
	16.50	57.8%	42.2%	100.0%
	18.50	53.3%	46.7%	100.0%
	20.50	58.6%	41.4%	100.0%
	22.50	57.7%	42.3%	100.0%
	49.50	48.2%	51.8%	100.0%
Total		75.0%	25.0%	100.0%

paredfour * highmath Crosstabulation

% within paredfour

		highmath		Total
		.00	1.00	
paredfour	1.00	87.3%	12.7%	100.0%
	2.00	83.9%	16.1%	100.0%
	3.00	67.1%	32.9%	100.0%
	4.00	52.9%	47.1%	100.0%
Total		75.0%	25.0%	100.0%

private * highmath Crosstabulation

% within private

		highmath		Total
		.00	1.00	
private	.00	77.8%	22.2%	100.0%
	1.00	61.2%	38.8%	100.0%
Total		75.0%	25.0%	100.0%

faminc * twopar Crosstabulation

% within twopar

		twopar		Total
		.00	1.00	
faminc	.75	14.0%	6.6%	9.8%
	2.50	22.2%	13.3%	17.1%
	4.50	19.1%	15.9%	17.2%
	6.50	13.3%	14.9%	14.3%
	8.50	10.0%	12.0%	11.1%
	10.50	6.4%	10.4%	8.7%
	12.50	4.6%	7.6%	6.3%
	14.50	3.2%	5.0%	4.2%
	16.50	1.7%	2.7%	2.3%
	18.50	1.2%	1.9%	1.6%
	20.50	1.5%	2.6%	2.1%
	22.50	0.4%	1.1%	0.8%
	49.50	2.3%	6.1%	4.5%
Total		100.0%	100.0%	100.0%

paredfour * twopar Crosstabulation

% within twopar

		twopar		Total
		.0000	1.0000	
paredfour	1.0000	43.1%	29.7%	35.5%
	2.0000	23.2%	18.8%	20.7%
	3.0000	20.6%	28.1%	24.8%
	4.0000	13.0%	23.5%	19.0%
Total		100.0%	100.0%	100.0%

private * twopar Crosstabulation

% within twopar

		twopar		Total
		.0000	1.0000	
private	.0000	89.4%	78.9%	83.4%
	1.0000	10.6%	21.1%	16.6%
Total		100.0%	100.0%	100.0%

Those living with two parents are more likely to score high in math than those not living with two parents. Those with higher family income are more likely to score high in math than those with lower family income. Those whose parents have more education are more likely to score high in math than those whose parents have less education. Those who attend private high school are more likely to score high in math than those who do not attend private high school.

Those living with two parents have higher family income than those not living with two parents. Those living with two parents have higher parental education than those not living with two parents. Those living with two parents are more likely to attend private high school than those not living with two parents.

Estimate models by using the logistic regression procedure with HIGHMATH as the dependent variable and the following independent variables:

model 1: TWOPAR
model 2: TWOPAR FAMINC TWOYR FOURYR GRAD
model 3: TWOPAR FAMINC TWOYR FOURYR GRAD PRIVATE

Variables in the Equation

		B	S.E.	Wald	df	Sig.	Exp(B)
Step 1[a]	twopar	.512	.034	227.314	1	.000	1.669
	Constant	-1.408	.027	2720.576	1	.000	.245

a. Variable(s) entered on step 1: twopar.

Variables in the Equation

		B	S.E.	Wald	df	Sig.	Exp(B)
	twopar	.228	.036	39.649	1	.000	1.256
	faminc	.021	.002	160.916	1	.000	1.021
Step 1[a]	twoyr	.231	.055	17.480	1	.000	1.260
	fouryr	1.052	.048	482.740	1	.000	2.864
	grad	1.543	.051	903.199	1	.000	4.676
	Constant	-2.150	.041	2763.553	1	.000	.116

a. Variable(s) entered on step 1: twopar, faminc, twoyr, fouryr, grad.

Variables in the Equation

		B	S.E.	Wald	df	Sig.	Exp(B)
Step 1[a]	twopar	.214	.036	34.625	1	.000	1.239
	faminc	.019	.002	124.796	1	.000	1.019
	twoyr	.226	.055	16.641	1	.000	1.253
	fouryr	1.025	.048	451.946	1	.000	2.787
	grad	1.509	.052	851.341	1	.000	4.523
	private	.242	.045	29.447	1	.000	1.274
	Constant	-2.151	.041	2765.686	1	.000	.116

a. Variable(s) entered on step 1: twopar, faminc_10, twoyr, fouryr, grad, private.

Controlling for family income and parental education explained 55% of the two-biological-parent effect found in the smallest model ((.228 − .512) × 100/.512) while controlling for private high school explained an additional 6% ((.214 − .228) × 100/.228). Controlling for all three variables explained 58% ((.214 − 512) × 100/.512).

Chapter 6. Modeling Interactions

1. Estimate models by using the logistic regression procedure with PRIVATE as the dependent variable and the following independent variables:

   model 1: BLACK OTHRACE PARCOLL
   model 2: BLACK OTHRACE PARCOLL BPARCOLL OPARCOLL
   model 3: BLACK OTHRACE WPARCOLL BPARCOLL OPARCOLL

Variables in the Equation

		B	S.E.	Wald	df	Sig.	Exp(B)
Step 1[a]	black	-.199	.071	7.899	1	.005	.820
	othrace	-.225	.044	26.136	1	.000	.799
	parcoll	1.400	.042	1119.089	1	.000	4.054
	Constant	-2.304	.038	3584.484	1	.000	.100

a. Variable(s) entered on step 1: black, othrace, parcoll.

Variables in the Equation

		B	S.E.	Wald	df	Sig.	Exp(B)
Step 1[a]	black	-.027	.108	.064	1	.800	.973
	othrace	-.146	.075	3.835	1	.050	.864
	parcoll	1.465	.055	705.419	1	.000	4.327
	bparcoll	-.285	.143	3.976	1	.046	.752
	oparcoll	-.117	.092	1.590	1	.207	.890
	Constant	-2.349	.046	2559.969	1	.000	.095

a. Variable(s) entered on step 1: bparcoll, oparcoll.

Variables in the Equation

		B	S.E.	Wald	df	Sig.	Exp(B)
Step 1[a]	black	-.027	.108	.064	1	.800	.973
	othrace	-.146	.075	3.835	1	.050	.864
	wparcoll	1.465	.055	705.419	1	.000	4.327
	bparcoll	1.180	.132	80.332	1	.000	3.255
	oparcoll	1.348	.074	329.823	1	.000	3.850
	Constant	-2.349	.046	2559.969	1	.000	.095

a. Variable(s) entered on step 1: black, othrace, wparcoll, bparcoll, oparcoll.

2. Estimate models by using the linear regression procedure with X2TXMTSCOR as the dependent variable and the following independent variables:

   model 1: BLACK OTHRACE PARCOLL
   model 2: BLACK OTHRACE PARCOLL BPARCOLL OPARCOLL
   model 3: BLACK OTHRACE WPARCOL BPARCOLL OPARCOLL

Coefficients[a]

Model		Unstandardized Coefficients		Standardized Coefficients	t	Sig.
		B	Std. Error	Beta		
1	(Constant)	49.065	.110		447.536	.000
	black	-4.535	.225	-.136	-20.115	.000
	othrace	.163	.145	.008	1.125	.261
	parcoll	6.799	.134	.334	50.680	.000

a. Dependent Variable: x2txmtscor

Coefficients[a]

Model		Unstandardized Coefficients		Standardized Coefficients	t	Sig.
		B	Std. Error	Beta		
1	(Constant)	49.087	.123		399.530	.000
	black	-3.720	.284	-.112	-13.095	.000
	othrace	-.161	.191	-.008	-.843	.399
	parcoll	6.754	.177	.332	38.230	.000
	bparcoll	-2.342	.467	-.043	-5.011	.000
	oparcoll	.821	.293	.027	2.801	.005

a. Dependent Variable: x2txmtscor

Coefficients[a]

Model		Unstandardized Coefficients		Standardized Coefficients	t	Sig.
		B	Std. Error	Beta		
1	(Constant)	49.087	.123		399.530	.000
	black	-3.720	.284	-.112	-13.095	.000
	othrace	-.161	.191	-.008	-.843	.399
	wparcoll	6.754	.177	.297	38.230	.000
	bparcoll	4.412	.433	.081	10.198	.000
	oparcoll	7.575	.234	.253	32.396	.000

a. Dependent Variable: x2txmtscor

Blacks have a lower effect for college-graduate parent than Whites, 4.412 compared with 6.754. The difference of 2.342 is significant at the .05 level with a *t* value of −5.011.

3. Estimate models by using the linear regression procedure with X2TXMTSCOR as the dependent variable and the following independent variables:

model 1: STEP SINGLE FAMOTH FAMINC
model 2: STEP SINGLE FAMOTH FAMINC STEPINC SINGLEINC FAMOTHINC
model 3: STEP SINGLE FAMOTH TWOINC STEPINC SINGLEINC FAMOTHINC

Coefficients[a]

Model		Unstandardized Coefficients		Standardized Coefficients	t	Sig.
		B	Std. Error	Beta		
1	(Constant)	50.240	.116		434.211	.000
	step	-1.989	.204	-.068	-9.733	.000
	single	-1.722	.178	-.069	-9.666	.000
	famoth	-2.749	.249	-.077	-11.031	.000
	faminc	.253	.007	.251	36.620	.000

a. Dependent Variable: x2txmtscor

Coefficients[a]

Model		Unstandardized Coefficients		Standardized Coefficients	t	Sig.
		B	Std. Error	Beta		
1	(Constant)	50.324	.124		405.682	.000
	step	-1.882	.278	-.065	-6.758	.000
	single	-2.136	.235	-.085	-9.102	.000
	famoth	-3.137	.326	-.088	-9.607	.000
	faminc	.245	.008	.243	30.121	.000
	stepinc	-.015	.022	-.007	-.687	.492
	singleinc	.059	.022	.025	2.725	.006
	famothinc	.056	.032	.016	1.759	.079

a. Dependent Variable: x2txmtscor

Coefficients[a]

Model		Unstandardized Coefficients		Standardized Coefficients	t	Sig.
		B	Std. Error	Beta		
1	(Constant)	50.324	.124		405.682	.000
	step	-1.882	.278	-.065	-6.758	.000
	single	-2.136	.235	-.085	-9.102	.000
	famoth	-3.137	.326	-.088	-9.607	.000
	twoinc	.245	.008	.239	30.121	.000
	stepinc	.229	.021	.099	11.068	.000
	singleinc	.304	.020	.128	15.050	.000
	famothinc	.300	.031	.085	9.820	.000

a. Dependent Variable: x2txmtscor

The effects of family income are equal for those living with two biological parents and those living with a biological parent/stepparent, .245 compared with .229, with the difference, $B = -.015$, not significantly different from zero with $t = -.687$). The effect of family income for those living with a single-parent family is more than the effect for those living with two biological parents, .304 compared with .245. The difference, $B = .059$, is significantly different from zero at the .05 level with $t = 2.725$.

Chapter 7. Modeling Linearity With Splines

1. Estimate models by using the linear regression procedure with X2TXMTSCOR as the dependent variable and the following independent variables:

   model 1: SESQ4_2 SESQ4_3 SESQ4_4
   model 2: SESQ4_2 JSESQ4_3 SESQ4_4
   model 3: JSESQ4_2 JSESQ4_3 SESQ4_4
   model 3: JSESQ4_2_SUM JSESQ4_3 SESQ4_4

Coefficients[a]

Model		Unstandardized Coefficients		Standardized Coefficients	t	Sig.
		B	Std. Error	Beta		
1	(Constant)	46.972	.133		354.019	.000
	sesq4_2	2.721	.186	.117	14.613	.000
	sesq4_3	5.314	.187	.228	28.475	.000
	sesq4_4	10.543	.187	.452	56.465	.000

a. Dependent Variable: x2txmtscor

Coefficients[a]

Model		Unstandardized Coefficients		Standardized Coefficients	t	Sig.
		B	Std. Error	Beta		
1	(Constant)	46.972	.133		354.019	.000
	sesq4_2	2.721	.186	.117	14.613	.000
	jsesq4_3	5.314	.187	.263	28.475	.000
	sesq4_4	5.229	.186	.224	28.161	.000

a. Dependent Variable: x2txmtscor

Coefficients[a]

Model		Unstandardized Coefficients		Standardized Coefficients	t	Sig.
		B	Std. Error	Beta		
1	(Constant)	46.972	.133		354.019	.000
	jsesq4_2	2.721	.186	.116	14.613	.000
	jsesq4_3	2.593	.185	.128	14.003	.000
	sesq4_4	5.229	.186	.224	28.161	.000

a. Dependent Variable: x2txmtscor

Coefficients[a]

Model		Unstandardized Coefficients		Standardized Coefficients	t	Sig.
		B	Std. Error	Beta		
1	(Constant)	46.972	.133		354.019	.000
	sesjsum	2.721	.186	.300	14.613	.000
	jsesq4_3	-.128	.321	-.006	-.399	.690
	sesq4_4	2.508	.263	.108	9.538	.000

a. Dependent Variable: x2txmtscor

2. Estimate models by using the linear regression procedure with X2TXMTSCOR as the dependent variable and the following independent variables:

model 1: X2SES
model 2: S01 S02 S03 S04 S05 S06
model 3: S01 S02 X2SES S04 S05 S06
model 4: S01 S02 S03 X2SES S05 S06
model 5: S_01_03 S_04_06
model 6: X2SES S_04_06

Coefficients[a]

Model		Unstandardized Coefficients		Standardized Coefficients	t	Sig.
		B	Std. Error	Beta		
1	(Constant)	51.138	.066		778.914	.000
	x2ses	5.455	.087	.403	62.397	.000

a. Dependent Variable: x2txmtscor

Coefficients[a]

Model		Unstandardized Coefficients		Standardized Coefficients	t	Sig.
		B	Std. Error	Beta		
1	(Constant)	43.845	.841		52.150	.000
	S01	2.908	1.424	.015	2.041	.041
	S02	4.722	.506	.089	9.327	.000
	S03	3.963	.380	.115	10.433	.000
	S04	7.322	.433	.191	16.897	.000
	S05	6.548	.652	.097	10.036	.000
	S06	5.602	1.959	.021	2.860	.004

a. Dependent Variable: x2txmtscor

Coefficients^a

Model		Unstandardized Coefficients		Standardized Coefficients	t	Sig.
		B	Std. Error	Beta		
1	(Constant)	50.780	1.030		49.305	.000
	S01	-1.055	1.422	-.006	-.742	.458
	S02	.759	.784	.014	.968	.333
	x2ses	3.963	.380	.292	10.433	.000
	S04	3.360	.730	.088	4.602	.000
	S05	2.585	.701	.038	3.690	.000
	S06	1.639	2.009	.006	.816	.415

a. Dependent Variable: x2txmtscor

Coefficients^a

Model		Unstandardized Coefficients		Standardized Coefficients	t	Sig.
		B	Std. Error	Beta		
1	(Constant)	56.660	1.144		49.521	.000
	S01	-4.415	1.504	-.023	-2.935	.003
	S02	-2.601	.615	-.049	-4.229	.000
	S03	-3.360	.730	-.098	-4.602	.000
	x2ses	7.322	.433	.540	16.897	.000
	S05	-.774	.972	-.012	-.797	.426
	S06	-1.721	1.948	-.007	-.883	.377

a. Dependent Variable: x2txmtscor

Coefficients^a

Model		Unstandardized Coefficients		Standardized Coefficients	t	Sig.
		B	Std. Error	Beta		
1	(Constant)	43.135	.254		169.892	.000
	S_01_03	4.277	.166	.195	25.727	.000
	S_04_06	6.897	.194	.269	35.573	.000

a. Dependent Variable: x2txmtscor

Coefficients^a

Model		Unstandardized Coefficients		Standardized Coefficients	t	Sig.
		B	Std. Error	Beta		
1	(Constant)	50.621	.090		560.468	.000
	x2ses	4.277	.166	.316	25.727	.000
	S_04_06	2.619	.315	.102	8.325	.000

a. Dependent Variable: x2txmtscor

3. Create four segment spline variables as follows:

COMPUTE S1=0.
COMPUTE S2=0.
COMPUTE S3=0.
COMPUTE S4=0.

IF -1.7501<=x2ses and x2ses<=-0.7501 S1=x2ses-(-1.7501).
IF x2SES>-.7501 S1=1.0000.

IF -0.7501<x2ses and x2ses<=0.2499 S2=x2ses-(-0.7501).
IF x2SES>0.2499 S2=1.0000.

IF 0.2499<x2ses and x2ses<= 1.2499 S3=x2ses-(0.2499).
IF x2SES> 1.2499 S3=1.0000.

IF 1.2499<x2ses and x2ses<= 2.2824 S4=x2ses-(1.2499).

COMPUTE S12 = S1 + S2.
COMPUTE S34 = S3 + S4.

COMPUTE check3= S1+S2+S3+S4+(-1.7501).

Descriptive Statistics

	N	Minimum	Maximum	Mean	Std. Deviation
x2ses	20133	-1.7501	2.2824	.090837	.7454149
check3	20133	-1.7501	2.2824	.090837	.7454149
Valid N (listwise)	20133				

Estimate models by using the linear regression procedure with X2TXMTSCOR as the dependent variable and the following independent variables:

model 1: S1 S2 S3 S4
model 2: X2SES S2 S3 S4
model 3: S12 S3 S4
model 4: X2SES S3 S4
model 5: S12 S34 S4
model 6: S12 S34

Coefficients[a]

Model		Unstandardized Coefficients		Standardized Coefficients	t	Sig.
		B	Std. Error	Beta		
1	(Constant)	43.373	.543		79.820	.000
	S1	3.999	.614	.048	6.511	.000
	S2	4.295	.240	.165	17.884	.000
	S3	7.063	.276	.238	25.589	.000
	S4	6.021	.832	.054	7.235	.000

a. Dependent Variable: x2txmtscor

Coefficients[a]

Model		Unstandardized Coefficients		Standardized Coefficients	t	Sig.
		B	Std. Error	Beta		
1	(Constant)	50.372	.562		89.648	.000
	x2ses	3.999	.614	.295	6.511	.000
	S2	.296	.758	.011	.390	.696
	S3	3.064	.636	.103	4.815	.000
	S4	2.021	1.051	.018	1.923	.054

a. Dependent Variable: x2txmtscor

The difference between the coefficient for S1 and S2 is .296 and not significantly different from zero. Thus, S1 and S2 can be combined.

Coefficients[a]

Model		Unstandardized Coefficients		Standardized Coefficients	t	Sig.
		B	Std. Error	Beta		
1	(Constant)	43.186	.258		167.360	.000
	S12	4.229	.172	.192	24.629	.000
	S34	7.095	.263	.277	26.946	.000
	S4	-1.098	.985	-.010	-1.115	.265

a. Dependent Variable: x2txmtscor

The next step is to see how the coefficients for S3 and S4 differ from the coefficient for S12.

Coefficients[a]

Model		Unstandardized Coefficients		Standardized Coefficients	t	Sig.
		B	Std. Error	Beta		
1	(Constant)	43.186	.258		167.360	.000
	S12	4.229	.172	.192	24.629	.000
	S3	7.095	.263	.239	26.946	.000
	S4	5.997	.830	.054	7.226	.000

a. Dependent Variable: x2txmtscor

The coefficients for S3 and S4 are significantly different from the coefficient for S12.

Coefficients[a]

Model		Unstandardized Coefficients		Standardized Coefficients	t	Sig.
		B	Std. Error	Beta		
1	(Constant)	50.588	.095		532.992	.000
	x2ses	4.229	.172	.312	24.629	.000
	S3	2.866	.385	.096	7.452	.000
	S4	1.768	.826	.016	2.140	.032

a. Dependent Variable: x2txmtscor

By adding S34, we can test whether the coefficients for S3 and S$ are equal. The difference between the coefficients for S3 and S4 is −1.098 and is not significantly different from zero.

Coefficients[a]

Model		Unstandardized Coefficients		Standardized Coefficients	t	Sig.
		B	Std. Error	Beta		
1	(Constant)	43.135	.254		169.892	.000
	S12	4.277	.166	.195	25.727	.000
	S34	6.897	.194	.269	35.573	.000

a. Dependent Variable: x2txmtscor

The final model includes two variables, one for S1 and S2 combined and one for S3 and S4 combined.

Chapter 8. Testing Research Hypotheses

1. Interval Independent/Interval Dependent

   The higher the *X*, the higher (lower) the *Y*.
   The higher the FAMILY INCOME,
   the higher the MATH SCORES.

2. Interval Independent/Dummy Dependent

   The higher the *X*, the more (less) likely to be *Y*.
   The higher the FAMILY INCOME,
   the more likely that the student attends PRIVATE HIGH SCHOOL.

3. Dummy Independent/Dummy Dependent

   Those in Category A are more likely to be *Y* than those in Category B.

   Those students in SINGLE-PARENT FAMILIES are more likely to attend PRIVATE HIGH SCHOOL than those not in SINGLE-PARENT FAMILIES.

4. Hypothesis for Analyzed Controls

   Part of the reason that the higher the *X*, the higher (lower) the *Y* is that the higher the *X*, the higher (lower) the *Z* and the higher the *Z*, the higher (lower) the *Y*.

   Part of the reason that those students who attend PRIVATE HIGH SCHOOL have higher MATH SCORES than those who do not attend PRIVATE HIGH SCHOOL is that those who attend PRIVATE HIGH SCHOOL have higher family income than those who do not attend PRIVATE HIGH SCHOOL and the higher the FAMILY INCOME, the higher the MATH SCORES.

5. Hypothesis for Interactions

   The effect of *X* on *Y* is larger (smaller) for Group A than for Group B.

   The effect of FAMILY INCOME on MATH SCORES is larger for those who attend PRIVATE HIGH SCHOOL than for those who do not attend PRIVATE HIGH SCHOOL.

References

Agresti, A., & Finlay, B. (2009). *Statistical methods for the social sciences.* Upper Saddle River, NJ: Prentice Hall.

Allen, M. P. (1997). *Understanding regression analysis.* New York, NY: Plenum Press.

Allison, P. D. (1999). *Multiple regression: A primer.* Thousand Oaks, CA: Pine Forge.

Demaris, A. (1992). *Logit models: Practical applications.* Newbury Park, CA: Sage.

Demaris, A. (2004). *Regression with social data: Modeling continuous and limited response variables.* New York, NY: Wiley.

Fox, J. (2015). *Applied regression analysis and generalized linear models.* Thousand Oaks, CA: Sage.

Frankfort-Nachmias, C., & Leon-Guerrero, A. (2014). *Social statistics for a diverse society.* Thousand Oaks, CA: Pine Forge.

Goldberger, A. (1998). *Introductory econometrics.* Cambridge MA: Harvard University Press.

Gordon, R. A. (2010). *Regression analysis for the social sciences.* New York, NY: Routledge.

Gordon, R. A. (2012). *Applied statistics for the social and health sciences.* New York: Routledge.

Greene, W. H. (2012). *Econometric analysis.* Boston, MA: Prentice Hall.

Hage, J. (1972). *Techniques and problems of theory construction in sociology.* New York, NY: Wiley.

Hardy, M. A. (1993). *Regression with dummy variables.* Newbury Park, CA: Sage.

Hill, R. C., Griffiths, W. E., & Judge, G. G. (1997). *Undergraduate econometrics.* New York, NY: Wiley.

Hosmer, D. W., Lemeshow, S., & Sturdivant, R. X. (2013). *Applied logistic regression.* New York, NY: Wiley.

Jaccard, J. (1990). *Interaction effects in multiple regression.* Newbury Park, CA: Sage.

Jaccard, J. (2001). *Interaction effects in logistic regression.* Thousand Oaks, CA: Sage.

Kahane, L. H. (2001). *Regression basics.* Thousand Oaks, CA: Sage.

Kutner, M. H., Nachtsheim, C. J., Neter, J., & Li, W. (2005). *Applied linear statistical models.* New York, NY: McGraw-Hill/Irwin.

Linneman, T. (2014). *Social statistics: Managing data, conducting analyses, presenting results.* New York, NY: Routledge.

Pampel, F. C. (2000). *Logistic regression: A primer.* Thousand Oaks, CA: Sage.

Pindyck, R. S., & Rubinfeld, D. L. (1998). *Econometric models and economic forecasts.* Boston, MA: McGraw-Hill.

Stinchcombe, A. L. (1968). *Constructing social theories.* New York, NY: Harcourt, Brace, & World.

Treiman, D. J. (2009). *Quantitative data analysis: Doing social research to test ideas.* New York, NY: Jossey-Bass.

Wooldridge, J. M. (2013). *Introductory econometrics: A modern approach.* New York, NY: South-Western.

Index